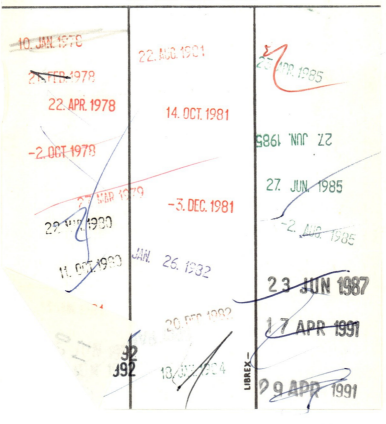

DEVELOPMENTS WITH
THERMOSETTING PLASTICS

*Proceedings of the symposium held at the
National College of Rubber Technology,
Polytechnic of North London*

DEVELOPMENTS with THERMOSETTING PLASTICS

Edited by

A. WHELAN

and

J. A. BRYDSON

National College of Rubber Technology, London

APPLIED SCIENCE PUBLISHERS LTD
LONDON

APPLIED SCIENCE PUBLISHERS LTD
RIPPLE ROAD, BARKING, ESSEX, ENGLAND

ISBN: 0 85334 612 7

WITH 85 ILLUSTRATIONS AND 45 TABLES

© APPLIED SCIENCE PUBLISHERS LTD 1975

All rights reserved. No part of this publication may be reproduced, stored in a retrieval system, or transmitted in any form or by any means, electronic, mechanical, photocopying, recording, or otherwise, without the prior written permission of the publishers, Applied Science Publishers Ltd, Ripple Road, Barking, Essex, England

Printed in Great Britain by Bell and Bain Ltd., Glasgow

Preface

The thermosetting plastics are a highly diverse, interesting and useful class of materials. The individual types show great versatility and may be used not only in such areas as moulding and laminating materials but also in cellular form, textile finishing, surface coating, adhesives and so on. Developments during the past few years are now coming in for intense scrutiny since the relative economics of many materials and processes have been so seriously affected by the vast increase in petroleum prices since the autumn of 1973.

It was with this in mind that at the National College of Rubber Technology, a constituent body of the Polytechnic of North London, we decided to organise a symposium considering such developments in the technological aspects of thermosetting materials. This may be considered a continuation of the policy of the NCRT to mount symposia on topics which are considered in need of review or appraisal. In recent years these have included PVC, polyurethanes, sealants, natural rubber and thermoplastics rubbers.

We were very fortunate with our selection of speakers, who not only provided well-delivered lectures of high standard but also gave us excellent cooperation both in the preparation of the symposium and of these published proceedings.

It will be seen that, following the introductory paper on prospects and trends for thermosetting plastics, the following five papers are concerned with different classes of material, specifically amino resins, polyesters, Friedel–Crafts resins, furane resins and polyurethanes. The final five lectures are more in the way of a review of aspects of the processing technology of these materials such as powder coating, polyurethane foam processing technology, FRP fabrication and the injection moulding of thermosets. This latter topic is now assuming such importance that two papers were given (Chapters 10 and 11), one by a machinery manufacturer and the other by a materials supplier. Throughout, the emphasis has been on the practical technology rather than the theoretical chemistry, since we believe that this is the area most neglected in the literature.

As editors we have tried to be consistent without, however, affecting the styles of the individual contributors. For example, whilst some authors preferred epoxy to epoxide when describing the resin, we have chosen the latter but retained the word epoxy to describe the chemical group. In many cases we have also provided data in both SI and f.p.s. units, the first set of figures in each case, normally, being the authors' choice. The second set are the equivalent to a degree of precision appropriate to the context. In cases where data were originally supplied in c.g.s. units we have given f.p.s. equivalents as first choice and SI second, deleting the original data.

A. WHELAN
J. A. BRYDSON

List of Contributors

A BARNATT
Technical Service Manager (Urethanes), Lankro Chemicals Ltd

J. B. BLACKWELL
Joint Chairman, Viking Engineering Co. Ltd

C. M. BROMLEY
Deputy Director, British Plastics Federation

R. A. BUTLER
Joint Managing Director, Butler-Smith Associates Ltd

R. H. CHAMBERS
Chief Chemist, Telcon Plastics Ltd

R. W. K. COOK
Technical Services Manager, Reinforcements Division, Fibreglass Ltd

A. G. EDWARDS
Senior Research Chemist, Industrial Chemicals Division, Albright and Wilson Ltd

F. J. PARKER
Group Leader, Moulding Materials Technical Service, British Industrial Plastics Ltd (Chemicals Division)

K. PARVIN
Senior Resin Chemist, Scott Bader Co. Ltd

A. T. RADCLIFFE
Technical Service and Development Chemist, Quaker Oats Ltd (Chemicals Division)

C. P. VALE
Research Manager (New Products), British Industrial Plastics Ltd (Chemicals Division)

Contents

Preface	v
List of Contributors	vii
1. Patterns of Usage of Thermosetting Plastics . . C. M. BROMLEY	1
2. Amino Resins . . . . C. P. VALE	13
3. Recent Developments in Unsaturated Polyester Resins . K. PARVIN	31
4. Friedel–Crafts Resins . . . . . . A. G. EDWARDS	41
5. Furane Resins . . . . A. T. RADCLIFFE	58
6. Recent Developments in Polyurethanes . . . A. BARNATT	76
7. Processing Equipment for Cellular Polyurethanes . . J. B. BLACKWELL	96

CONTENTS

8. **Powder Processing of Thermosetting Plastics** . . 110
   R. H. CHAMBERS

9. **FRP Low Pressure Processes** 127
   R. W. K. COOK

10. **Machinery for Thermoset Injection Moulding** . . 149
    R. A. BUTLER

11. **Injection Moulding—Material and Process Requirements** 162
    F. J. PARKER

Index 189

CHAPTER 1

Patterns of Usage of Thermosetting Plastics

C. M. BROMLEY
(*British Plastics Federation*)

1.1 INTRODUCTION

This paper examines the development of the thermosetting plastics industry up to the present time, gives the present pattern of usage and surveys prospects up to 1980.

1.2 DEVELOPMENT OF THE THERMOSETTING PLASTICS INDUSTRY

1.2.1 Developments up to 1968

The production of synthetic plastics materials started nearly 70 years ago with phenol-formaldehyde (PF), followed by urea-formaldehyde (UF) in 1926 and melamine-formaldehyde (MF) in 1935.

Whilst the early 1930s and World War II saw the development of the newer family of thermoplastics, thermosetting plastics were still the most important until the early 1950s, as Table 1.1 shows. Thus,

TABLE 1.1
UK PRODUCTION OF PLASTICS MATERIALS, 1951–68 (1000 TONNES)

	1951	*1953*	*1956*	*1959*	*1965*	*1968*
Thermosetting plastics	130	110	158	235	312	330
Thermoplastics	65	100	155	423	630	855
Total	195	210	313	658	942	1185

Source: NEDO.

it may be seen that thermosetting plastics accounted for 66% of UK production in 1951 but by 1968 their share had fallen to 28%.

Of the total production of 330 000 tonnes in 1968, 73% was accounted for by the older materials, PFs, UFs, MFs and alkyds. The remaining 27% was accounted for by the newer materials such as unsaturated polyesters (UP), epoxides (EP) and polyurethanes (PU).

1.2.2 Developments from 1968 to 1972

The UK production of thermosetting plastics over this period has developed as shown in Table 1.2.

TABLE 1.2

UK PRODUCTION OF THERMOSETTING PLASTICS, 1968–72 (1000 TONNES)

	1968	1969	1970	1971	1972
Aminoplasts	130·1	136·3	137·1	137·8	143·0
Phenolics and cresylics	68·4	73·5	59·9	55·3	54·5
Alkyds	65·0	64·3	66·4	71·3	60·6
Unsaturated polyester resins	34·0	38·5	40·8	39·4	47·9
Polyurethanes (flexible and semi-flexible)	35·3	30·9	33·4	36·4	41·8
Other thermosets	23·5	33·1	38·8	37·4	38·2
Total	356·3	376·6	376·4	377·6	386·0

Source: BPF Statistical Handbook.

For this period, i.e. 1968–72, this gives an annual growth rate of 2·4% for aminoplasts, and an annual decline of 5·5% a year for phenolics and cresylics and of 1·7% a year for alkyd resins. By contrast, the output of polyester resins rose by an average of 8·9%, that of polyurethanes by 4·3% and of the other thermosetting plastics by no less than 12·9%. Overall, the growth of thermosetting plastics output was only 1·9% a year. The effect of these different growth rates can be seen in Table 1.3.

During this time, the trend in consumption was as given in Table 1.4. These estimates have been prepared by taking the figures for production, subtracting exports and adding imports.

1.2.3 Subsequent developments

As is now generally known, 1973 was a year of considerable expansion in demand and output. According to *Europlastics Monthly*, the estimated figures are as shown in Table 1.5.

TABLE 1.3
PERCENTAGE SHARE OF PRODUCTION OF THERMOSETTING RESINS, 1968–72

	1968	1970	1972
Aminoplasts	36·3	36·4	37·1
Phenolics and cresylics	19·1	15·9	14·1
Alkyds	18·1	17·6	15·7
Unsaturated polyester resins	9·5	10·8	12·4
Polyurethanes	9·9	8·9	10·8
Others	7·1	10·4	9·9
Total	100·0	100·0	100·0

TABLE 1.4
CONSUMPTION OF THERMOSETTING PLASTICS, 1968–72 (1000 TONNES)

	1968	1969	1970	1971	1972
Aminoplasts	110·8	116·8	111·3	110·7	110·9
Phenolics and cresylics	54·2	51·7	39·8	33·9	42·4
Alkyd resins	55·9	53·8	57·6	63·1	54·5
Unsaturated polyester resins	33·4	37·2	40·9	38·5	48·0
Polyurethane foams	35·7	30·5	33·1	36·2	41·3
Other thermosets	n/a	n/a	40·0	37·5	34·8
Total	290·0	290·0	322·7	319·9	331·9

TABLE 1.5
PRODUCTION AND CONSUMPTION OF THERMOSETTING PLASTICS, 1973 (1000 TONNES)

	Production	Consumption
Aminoplasts	155	133
Phenolics	56	46
Alkyds	66	38
Unsaturated polyesters	52	55
Polyurethanes	72	70
Epoxides	16	13
Others	15	25
Total	432	380

Aminoplastics—The demand for aminoplastics was high from the electrical and consumer durable sectors and was also influenced by a shortage of phenolic moulding compounds.

Europlastics have stressed that this material is not wholly dependent on oil. Therefore, although aminoplastics have lost out to thermoplastics the availability of the latter could change future demand patterns.

TABLE 1.6

ESTIMATED PATTERN OF USAGE OF AMINOPLASTS, 1972 (1000 TONNES)

Moulding powders	45·9
Adhesives	44·0
Surface coatings	10·5
Textiles	4·0
Paper treatment	5·5
Laminates	8·9
Others	11·2
Total	130·0

Source: Europlastics Monthly.

Phenolics—The major growth areas were in the motor and electrical industries. In the second half of 1973, restricted supplies of phenol reduced the use of phenolic materials. During 1974/75, the shortage of phenol will persist and this will restrict the applications for this class of materials.

Unsaturated polyesters—Major progress in 1973 was in the building industry, particularly for cladding. Road, air and marine transport were also good growth areas.

Epoxides—The major growth areas were surface coatings, powder coatings and corrosion resistant applications.

Polyurethanes—Consumption was held back due to materials shortages, particularly MDI, TDI and polyols. For flexible foams, the major areas of application were furniture and bedding. In packaging, PU encountered stronger competition from expanded polystyrene (EPS).

As the domestic refrigerator and deep freeze business experienced considerable growth the usage of rigid foams also increased.

1.2.4 Application areas for individual materials

In this area a less complete picture exists. The data have been built up from various sources, notably the NEDO Report and *Europlastics*

TABLE 1.7
ESTIMATED PATTERN OF USAGE OF PF, 1969 (1000 TONNES)

Moulding materials	14·1
Laminates	14·7
Bonding and adhesives	10·3
Friction materials	7·7
Foundry applications	13·3
Coating and varnishes	5·4
Other applications	5·0
Total	70·5

TABLE 1.8
ESTIMATED PATTERN OF USAGE OF UNSATURATED POLYESTER RESINS, 1972 (1000 TONNES)

Transport	7·3
Body fillers	3·7
Building	7·8
Marine	6·8
Pipe seals	3·0
Chemical plant	1·5
Electrical	1·6
Insulated containers	2·1
Others	3·5
Total	37·3

Monthly. The first materials considered are the aminoplasts and the pattern of usage is shown in Table 1.6. Table 1.7 indicates the pattern of usage for the phenolics.

As for alkyd resins, decorative and industrial paints are of equal importance—amounting together to nearly 90% of usage. The remainder is accounted for by marine paints. For the newer materials the patterns of usage are set out in Tables 1.8–1.10.

TABLE 1.9
ESTIMATED PATTERN OF USAGE FOR EPOXIDES, 1972 (1000 TONNES)

Solid resins	
Surface coatings	5·0
Electrical	1·2
Liquid resins	
Surface coatings	1·2
Electrical	1·0
Tooling	0·2
Adhesives	0·5
Laminates	0·3
Flooring	0·6
Miscellaneous	0·5
Total	10·5

TABLE 1.10
ESTIMATED PATTERN OF USAGE OF PU, 1970 (1000 TONNES)

Flexible and semi-flexible foam	
Transport	10·0
Furniture	12·5
Bedding	3·6
Textile foam backs	6·1
Packaging	1·5
Household	5·8
Miscellaneous	0·5
Total	40·0
Rigid foam	
Refrigerators and freezers	2·0
Cold stores	0·8
Insulated transport	1·2
Building	1·0
Miscellaneous	2·0
Total	7·0
Combined total	47·0

1.3 FORECASTS OF PRODUCTION AND CONSUMPTION OF THERMOSETTING MATERIALS

Having discussed and reviewed the pattern of usage in 1972 and 1973, it is interesting to compare the actual outcome with what was forecast in the NEDO Report of 1972 (*The Plastics Industry and its Prospects*, London, HM Stationery Office, 1972) (Table 1.11). These estimates

TABLE 1.11

FORECASTS OF PRODUCTION OF THERMOSETTING MATERIALS 1968–80 (1000 TONNES)

	1968	1980	% a year 1968–80
UF and MF	92	115	2
PF	68	110	4
Alkyds	80	100	2
UP	34	100	9
EP	8	25	10
PU	43	145	11
Others	5	20	12
Total	330	615	5

Source: NEDO.

confirmed the historical trends so that, by 1980, the three groups of older materials would account for 51% of output, growing at only 3% a year. The newer materials were expected to grow at 10% a year (Table 1.12).

The market breakdown for 1980 was forecast as given in Table 1.13. If these estimates are looked at in detail the dependence of the thermosetting plastics on the building, automotive and furniture markets may be clearly seen. For building, the main areas of application are for binders and adhesives, surface coatings and insulation. In the areas of binders and adhesives one important application is chipboard. The synthetic product is in direct competition with wood and, so far, synthetic materials have retained their competitive position due to shortages of the natural commodities. For insulation, there is inter-plastics competition as well as strong competition from other materials. There are stronger economic pressures for insulation than hitherto and yet, at the same time, there are indications of tougher attitudes by specifying authorities and insurers.

TABLE 1.12
FORECASTS OF CONSUMPTION OF THERMOSETTING PLASTICS (1000 TONNES)

	1968	1980	% a year 1968–1980
UF and MF	83	110	2
PF	65	100	4
Alkyds	71	90	2
UP	34	100	9
EP	8	24	10
PU	42	140	11
Others	5	20	12
Total	308	584	5

TABLE 1.13
ESTIMATED PATTERN OF USAGE OF THERMOSETTING PLASTICS, 1980 (1000 TONNES)

	UF/MF	PF	Alkyds	UP	EP	PU
Building	63·2	35·0	20·0	29·1	7·0	35·0
Packaging	—	—	—	—	—	—
Electrical products	8·6	26·0	—	0·4	10·0	1·9
Automotive products	1·0	3·5	—	36·0	—	52·0
Furniture	33·5	14·0	—	5·5	—	42·0
Others	4·0	21·5	70·0	22·0	4·5	9·0
Total	110·3	100·0	90·0	93·0	21·5	139·9

As to automotive products, the industry's output has fluctuated widely since 1972, having fallen from the 1968–71 period. The industry's future investment and its usage of materials will thus be affected.

For furniture, existing applications have been consolidated and expanded. It still remains a major area for expansion.

1.4 MAJOR ASSUMPTIONS IN THE NEDO FORECASTS

These can be summarised as follows: a steady rate of growth in the economy, averaging 3% a year from 1968–80; two-thirds of the annual average growth in plastics consumption would come from innovation; an adequate supply of materials from UK sources; plastics remaining broadly price competitive with other materials.

1.4.1 The UK plastics industry and the UK economy

This was a subject I covered at the recent conference 'The Challenge Facing the UK Plastics Industry, 1974 and Beyond', and I make no apology for returning to it again.

The NEDO report indicated that the UK's usage of plastics had been restricted due to the relatively slow growth of the UK economy. This latter factor has also made the UK a relatively unattractive investment location.

From the report's base year of 1968, the UK economy only grew by an average of just under 2% a year. As a result, the UK demand for plastics grew more slowly than anticipated. Over-capacity resulted, with an accompanying weak price structure, particularly for the newer thermosetting plastics. By 1971, the materials producing industry was experiencing a combination of circumstances which produced unprofitability. Demand began to recover in 1972 and boomed in 1973 as the UK achieved an annual growth rate of nearly 6%.

This tremendous expansion of the UK economy began to suffer difficulties as early as the middle of 1973. These difficulties increased during the second half of 1973 and were accentuated by the Arab oil policy, followed by the Middle Eastern war and by the UK's industrial problems.

From initial predictions of no growth in the economy for 1974, the position is now reached whereby most forecasters expect there to be a net decrease in the UK's Gross National Product from 1973 to 1974 and this may persist through to the first part of 1975. It would be both emotive and inaccurate to describe this as a recession but it should be seen as a large fluctuation in the cycle of growth.

These poor short-term prospects are significant because of their relationship to the need of the UK plastics industry and of its thermosetting plastics sector. These needs have to be seen in the context of the UK's basic economic situation. The various forecasts from OECD and the NIESR identify three main trends: a decline in the UK's GDP growth rate; a deteriorating balance of payments; an acceleration in the rate of inflation.

There are a number of comments to make about these three main trends. First, the UK plastics industry becomes a more significant part of the UK's economy each year, as its growth rate is so much faster than that of the UK economy. Secondly, the UK's balance of trade has deteriorated sharply not simply because of increased oil and other commodity prices. Thirdly, the UK plastics industry has a substantial record of containing inflation in the past, as the prices of materials have either stayed constant or fallen in real terms. These price reductions have come more from the materials producers rather than the processors.

As the industry knows, 1973 saw the development of serious shortages due in part to a lack of materials production capacity. There is a need to invest, as this will lead to: increased supplies of materials and processing capacity as well as increased capacity in the downstream customer industries; a more competitive economy, embracing the various customer industries as well as the thermosetting industry, leading to an improved balance of trade and favourable currency rates; the containment of inflation through better unit costs.

The UK plastics industry thus needs investment in the customer industries as well as investment itself. The dilemma is that investment depends on demand prospects whilst at the same time customers may have to base their own investment decisions on their assessment of suppliers.

1.4.2 Innovation

This topic covers the areas of marketing and technical development and the allocation of company resources to developing new business. There are a number of aspects of this definition.

Marketing
Companies will only be able to develop their business if they have a proper base of market information not only about their present markets but also about future business prospects. Marketing should not be seen as being related only to market data but should also include financial information about the products.

Technical development

This may cover several headings. For example, the improvement of existing products, the improvement of existing production methods, and the development of new applications and products. It has to be stressed that this technical development, wherever it might be undertaken, has to be financed and must be related to new business and the improvement of existing business.

Finance

A substantial increase in capacity utilisation in the materials and processing sectors occurred in 1973. Despite the statutory price controls there was an improvement in profitability. However, the effects of increased material prices and shortages has significantly affected companies' working capital requirements and cost flow. This is most marked in stock levels and credit control.

Manpower

This is the final ingredient in any process of innovation and yet it is the most important. The interesting thing about 1973 is the clear suggestion that even if all the materials had been available there would not have been the manpower to process them. (As a first estimate the total employment in processing might be 40 000–50 000.)

Structure

The production of thermosetting resins is a smaller scale, less capital-intensive process than thermoplastics. For some materials such as alkyds, there are a considerable number of producers. As to the processing of materials, this is in some cases a smaller scale operation except for polyurethanes.

1.4.3 Supply of materials

As indicated earlier, the UK has experienced a shortage of materials due to a lack of investment. This situation will in general persist for the immediate future. This does have several implications: (a) companies need to achieve a more effective use of materials by improving the product design and the processing operations; (b) companies will need to make a number of strategic decisions about product development and sales; (c) companies will have to assess the extent to which restricted supplies of materials in the short–medium term will curtail market development.

1.4.4 Plastics and other materials

The change in plastics prices must be known to you all. The change was particularly marked in the UK, where the price structure of all materials was unreal due to over-capacity and a slowly-growing demand until 1971. Recovery came in 1972 and 1973 at a time when price controls were imposed. However, substantially increased feedstock and capital costs led to increased material prices. Shortages of materials also produced commodity speculation in spot lots.

These increases, which occurred with varying intensity across the whole range of materials, led in turn to increases in product prices. Initially this produced relatively little effect on demand. This was hardly surprising in view of the shortages of alternative materials and the capacity for processing them. Companies should be careful in the medium–longer term, however, of assuming that customers and users will continue to pay the higher prices. It will be necessary for the industry to look at its cost structure, to review the materials content of the final selling price and to press for higher added value products.

In some sectors at least the trend must be away from volume items to speciality performance products, with the accompanying quality and performance innovations.

1.5 CONCLUSIONS

It must be emphasised that despite the immediate short-term, gloomy national economic situation, there are substantial growth prospects for thermosetting plastics. The development of these growth prospects requires the determination of the industry to invest, to plan and to improve its management. It is, however, absolutely necessary for the industry to achieve a proper development strategy, avoiding the pitfalls of gloom (and the consequent reluctance to invest) on the one hand and technologically based optimism on the other. The industry's strategy must be confidence based on realism and professionalism.

The pattern of usage is likely to be affected initially by supply factors and both the industry's and the customers' response to these factors. In the longer term, the pattern of usage will be affected by the success or failure of the industry's own efforts.

CHAPTER 2

Amino Resins

C. P. VALE

(*British Industrial Plastics Ltd*)

2.1 STATISTICAL REVIEW

Figures given for the production of amino resins by different countries are invariably difficult to interpret. Amino resins for different purposes are manufactured and sold as solutions (in water or organic solvents) containing between 60 and 85% dry resin, or as solid (i.e. 100%) resin powders. Amino moulding powders normally contain about 70% by weight of dry resin, the remainder being cellulose filler and small amounts of additives such as pigments, accelerators, and mould lubricants. UK figures for *amino resin* production supplied by the Board of Trade (see Table 2.1) are understood to comprise the actual weight of resins (including solvent) plus the resin content of moulding powders.

In Table 2.2 aminoplastics production is shown in relationship to that of other plastics materials[1] for the year 1972.

From Table 2.1 it can be shown that over the five-year period 1967–72 the compound growth rate of amino resin production in the UK was 4·2% per annum, while that of amino moulding powders was 3·6%. The growth rate during the same period for all thermosetting plastics was 3·5% and for all plastics 8·0%. However, in spite of this relatively slow growth, the aminoplasts in terms of total production rank first among thermosets and fourth among all plastics types in this country.

Total world production is difficult to assess because some countries such as the USA give production figures in terms of resin

TABLE 2.1
UK NET PRODUCTION AMINOPLASTICS (1000 TONS)

Year	Amino resins	Amino moulding powders
1967	116·6	29·4
1968	128·0	31·6
1969	134·1	32·5
1970	137·1	31·3
1971	137·8	34·6
1972	143·1	35·0

TABLE 2.2
UK NET PRODUCTION OF PLASTICS MATERIALS, 1972 (1000 TONS)

Thermosetting plastics		Thermoplastics	
Aminoplasts	143·1	Polyethylene (L.D. and H.D.)	434·2
Alkyds (net resin content)	60·5	PVC and copolymers	344·3
Phenolics and cresylics	54·5	Styrene polymers and copolymers	183·9
Unsaturated polyesters	44·5	Polypropylene	117·2
Polyurethanes (flexible)	41·8	Polyvinyl acetate (net resin content)	44·4
Epoxides (net resin content)	14·8	Cellulose esters and ethers	9·7
Other (casein, rigid PU etc.)	23·3	Others (acrylics, nylons, PTFE, etc.)	85·9

TABLE 2.3
UK USAGE IN DIFFERENT APPLICATION FIELDS (1973)

Usage	Percentage
Moulding materials	43
Adhesives	22
Coating resins	11
Textile resins	5
Paper resins	5
Laminating resins	4
Others—including foam	10

dry weight, and in other cases the manner of compiling statistics of production is not made clear. It is estimated, however, that the total world production of aminoplastics, on the same basis as the UK figures of Table 2.1, will, in 1974, exceed 3m tons or, on dry weight basis, be between 2m and 2·5m tons. The leading manufacturing countries are Japan, USA, West Germany, the Soviet Union, Italy, France and the UK, probably in that order.

Currently, UK production is estimated[2] to be divided among the various fields of application, as shown in Table 2.3.

2.2 RAW MATERIALS

The three principal raw materials used in the manufacture of amino resins are formaldehyde, urea and melamine, and some interesting developments affecting the production of each of these three chemicals have taken place during the past few years. ICI Ltd have developed a new and more economical method for manufacturing methanol from which formaldehyde is usually made by catalytic dehydrogenation. The ICI process involves the use of a special copper catalyst which permits effective production at lower temperatures and considerably lower pressures than has hitherto been possible. The reaction involved is

$$CO + 2H_2 \rightleftharpoons CH_3OH$$

the appropriate carbon monoxide and hydrogen mixture usually being obtained by passing natural gas (mainly methane), or refinery gases, with steam over a heated catalyst.[3]

One of the long-recognised problems in the manufacture of urea by reacting carbon dioxide and ammonia under pressure is the highly corrosive nature of the reactant mixture at elevated temperatures. This tends to limit the temperature at which the process can be operated and thus the degree of conversion. Silver- or lead-lined autoclaves have been employed in the past, but more recently stainless steel passivated by incorporating oxygen in the reactant stream has been found to be more suitable, allowing somewhat higher operating temperatures. This and other process developments have resulted in improved degrees of conversion and a major reduction in production costs.[4]

Within the past five or six years there has been a very considerable movement towards manufacturing melamine from urea, the overall reaction being

$$6\ CO(NH_2)_2 \rightarrow C_3H_6N_6 + 3CO_2 + 6NH_3$$

After many years of research, satisfactory processes appear to have been developed by DSM, and BASF in Europe and Allied Chemicals in the USA, and most plants are now operating one of these. The dicyandiamide route, however, is not completely obsolete and at least three manufacturers, including BOC in the UK, are still making melamine by this method.

TABLE 2.4

APPLICATIONS OF AMINO MOULDING MATERIALS

Application	Percentage
Electrical (plugs, sockets, etc.)	49
Domestic appliances	10
Closures	23
Lavatory seats	6
Tableware	5
Miscellaneous	7

2.3 AMINO MOULDING MATERIALS

The most important advance is this field in recent years has been the development of grades of moulding material amenable to injection moulding. This topic is the subject of a later chapter and need not be considered further here. A few items of general information on amino powders, however, may usefully be mentioned.

An estimate of the extent to which aminos are currently employed in different fields is shown in Table 2.4.

Cellulose-filled UF materials form by far the greater part of the market, cellulose-filled MF (between 5 and 10% of total production) being used mainly in the production of tableware. Small quantities of asbestos-filled and glass-filled MF materials have been supplied in the past for special electrical purposes, but the demand has never been high and production of such materials in the UK has now ceased.

Both UF and MF moulding materials are invariably made by what is sometimes described as the 'wet processing method'; i.e. cellulose filler is mixed with an aqueous solution of the UF or MF constituent, the mixture is then oven dried, ball-milled with pigments, mould lubricants and other additives and then densified, granulated or pelleted. Moulding materials can be produced by blending dry resin and filler, and compounding on a heated two-roll mill or in a Buss Ko-Kneader, but this method is not employed on a commercial scale.

2.4 SELECTED DEVELOPMENTS IN AMINO RESINS

The versatility of the amino resins is well recognised. In addition to their use in moulding compositions and in the manufacture of decorative laminates, they continue to be employed extensively as adhesives in the production of plywood and particle board, as agents in numerous textile finishing processes, as 'wet end' additives for imparting wet strength to paper, and as constituents of stoving enamels for cars, refrigerators, washing machines, etc. In this paper three topics, in which interesting advances have been made in recent years, have been selected for a more detailed consideration.

2.4.1 Urea-formaldehyde foams

UF foams may be regarded as the earliest of all commercial plastics foams. They were developed in Germany in the early part of World War II as heat insulating materials, under the trade-name 'Iporka',[5] but attained little commercial importance owing to their poor mechanical properties. Interest, however, was aroused in the late 1950s in the possibility of employing them as thermal insulators for dwelling houses[6] and, after an initial period of hesitancy and scepticism in this country, the use of UF foams for this purpose is growing rapidly.

Modern dwelling houses are constructed with a double wall comprising a brick outer leaf and an inner leaf of bricks or lightweight blocks. Between the two is a cavity of about 5 cm (2 in) width. The cavity reduces the risk of damp penetration and to a certain extent provides a small measure of thermal insulation, since air is a good insulator. However, air movement by convection within the cavity tends to transfer a considerable amount of heat from the inner to the outer leaf. This movement can be virtually eliminated by filling the cavity with a lightweight, rigid UF foam. By so doing the heat flow from a cavity wall can be reduced from about 1·7 to 0·5W m^{-2} $°C^{-1}$ or even less (i.e. down from 5·61 to 1·7 or less Btu in ft^{-2} h^{-1} $°F^{-1}$).

UF resins used for cavity wall insulation are similar to those employed as adhesives. They are made by condensing urea and formalin at the boil under slightly acidic conditions until the syrup obtained has achieved a required degree of hydrophobicity. It is then neutralised and water removed by distillation until a product of appropriate solids content, viscosity and stability is achieved. The

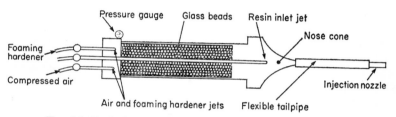

FIG. 2.1. Typical arrangement of construction of foaming gun.

resin can be diluted with water to a certain extent, but too great a water addition results in the separation of the resin. A solution of about 25% resin concentration is normally satisfactory for cavity wall foaming.

The other ingredient employed is a dilute solution of a mineral acid (usually phosphoric) containing also a foam-inducing surface-active agent.

Foaming is carried out *in situ*, using a specially constructed gun (see Fig. 2.1) which is fed with resin solution and foaming hardener from two separate pressurised pots. The hardener solution is forced with compressed air over a packed column of glass spheres forming a fine foam. The resin solution is injected into this foam which, after leaving the gun nozzle, gels within about 45 sec. The hardener is carefully formulated to ensure relatively rapid gelation before the foam has time to collapse. Houses already erected can be insulated in this way by boring holes through the mortar of the outer leaf at recommended intervals, filling the cavity by applying the foam gun at these points and ultimately sealing the holes with matching mortar.

Houses in course of erection are preferably treated through the inner leaf and prior to plastering.

Urea-formaldehyde foam is cheap, non-inflammable and hydrophobic, although the dry foam is permeable to water vapour. The foam used in cavity walls has a density of about 8 $kg\,m^{-3}$ (0·5 lb ft^{-3}). It has poor compressive strength but is self-supporting. The average detached house can be insulated with UF foam in one day by two men at a cost which has been estimated to be covered in about five years by savings in fuel bills. Tests have indicated that UF foam, once installed, is likely to last as long as the building itself.

Denser and stronger foams can be produced by using more concentrated resin solutions and making adjustments in the foam hardener. Several machines have been built capable of producing UF foam of a higher density.

Other uses found for UF foam of higher density include blocks for flower arrangements and artificial snow film effects. A recent report has mentioned that Plastic Techniques Ltd have developed a machine for the continuous production of UF foam insulating board, having a density of 24 $kg\,m^{-3}$ (1·5 lb ft^{-3}).[7]

The most interesting potential application of UF foams, however, is their use in the construction of arrester beds at the end of airfield runways. The purpose of these beds is to bring a plane rapidly and safely to a halt if, for any reason, it should overrun the runway. A development programme sponsored by the Civil Aviation Authority, BIP Ltd and the Royal Aircraft Establishment has led to trials with a Comet 3B aircraft. The latter, fully fuelled and weighing 55 tons has run into a bed of UF foam 90 m (295 ft) long and 0·6 m (2 ft) deep at a speed of 87 $km\,h^{-1}$ (47 knots) and been brought to a halt after travelling 55·5 m (180 ft).

The optimum structure for a UF foam bed used for this purpose consists of a lower layer of dense foam with a top layer of lighter foam. The width of the bed is equal to that of the runway and the end of the bed is ramped to facilitate smooth entry. Mean retardation is about 0·54g (where g is the acceleration due to gravity) and is independent of the speed of entry. Beds 180–300 m (590–984 ft) long could safely arrest any type of airliner, including Jumbo Jets and Concorde. The foam is crushed by the aircraft wheels but the foam particles do not damage the aircraft, even when ingested into the engines. In this way UF foam is superior to gravel or sintered fuel ash, which have been considered and tested for this application.[8]

2.4.2 Textile finishing processes

It has been known since the mid-1920s that the tendency of cellulosic fabrics to crease when folded or crumpled can be substantially obviated by treatment with amino resins. The process consists of padding the fabric through an aqueous solution of methylol ureas or methylol melamines (or preferably their more soluble and more stable methyl ethers) containing also an acid or acid-liberating

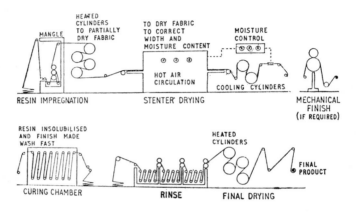

FIG. 2.2. Resin finishing of textiles.

catalyst. Excess solution is removed, and uniform impregnation ensured, by passing through mangle rolls. Thereafter, the fabric is dried on a stenter with hot air and finally passed through a chamber at 140–160°C to cure the amino resin (Fig. 2.2). It is preferable for the fabric to be given an alkaline wash to remove acid catalyst and any uncured resin, but this step is nearly always omitted.

It is desirable that the amino resin should be initially of low molecular weight to allow complete penetration of the cellulose fibres. More highly condensed amino resins adhere to the surface of the fibres and, after curing, produce a fabric with a stiff handle. The amount of resin pick-up depends on the concentration of amino component in the padding bath and the pressure exerted on the mangle rolls. Satisfactory crease resistance, and dimensional stability on saturating with water, is obtained with a resin pick-up of between 5 and 10% in the case of cotton and between 10 and 15% in the case of rayon. The permanence of these effects to repeated launderings depends on the water resistance of the cured resin and is therefore considerably greater with melamine resins than with those made from urea.

Other useful finishing processes, highly durable to washing and dry cleaning, have been developed using amino resins. Water-repellent finishes, for example, are obtained by padding cellulose fabrics through solutions containing a low molecular weight melamine resin and an aqueous dispersion or emulsion of stearamide or N-methylol stearamide. Glaze effects and embossed patterns can also be rendered permanent to laundering by impregnating the fabric with a solution of a melamine resin precondensate and catalyst, drying, and then calendering and finally baking at about 140°C for a few minutes to effect cure. Amino resins have also been used in the production of durable flameproof finishes on cotton fabrics; in pigment pastes for producing washproof and non-fading prints; and, in small quantities, for improving the wash-fastness of certain direct cotton dyes. Various processes involving the use of amino resins have been developed for rendering cellulose fabrics rotproof. More highly condensed UF resins have been used alone, or in conjunction with starch, for imparting permanent stiffening effects to cotton fabrics, and amino resins have been used in the production of durably stiffened nylon dress net.

In recent years growing competition from the rapidly developing synthetic fibres has prompted American cotton growers to finance an intense programme of research with the object of developing finishes allowing cotton fabrics to match the properties of the synthetics. The immense amount of work carried out in this field may be illustrated by the fact that between 1965 and 1973 over 9000 patents and 4000 literature references concerned with textile finishing processes have been noted.[9]

Fabrics made from synthetic fibres have the property, after washing, of drying to an almost creaseless condition. Research was directed, therefore, to producing a similar property in cotton. It was found that this could be achieved by applying to cotton fabrics rather larger amounts of amino resins than was required to give satisfactory dry crease resistance. The fabrics were then found to possess a measure of wet crease resistance. That is to say, on drying they required little or no ironing to give a smooth finish. Such cottons are now well known by the term 'drip dry', although the preferred term is 'smooth drying' or 'self smoothing'.

In the early development of such finishes two major problems were encountered. First the application of amino resins to cotton fabrics in the manner described above is always accompanied by a reduction in tensile strength and in resistance to abrasion. The fall in tensile strength is directly proportional to crease recovery and becomes quite serious where fabrics have been treated to obtain smooth drying effects (Fig. 2.3). Various processing techniques have been developed which to a greater or lesser extent overcome the difficulty

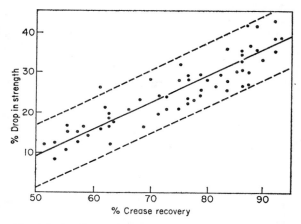

Fig. 2.3. Tensile strength and crease recovery for cotton.

of a reduction in the mechanical strength of the fabric. These need not be considered here. Few of them have been wholly successful on a commercial scale and the general trend now is to avoid the problem by using fabrics comprising blends of cotton and polyester fibres.

The second problem met with as a result of resin treatment arises from the use (more particularly in the USA) of hypochlorite bleaches in laundering processes. Where the fabric has been treated with a UF resin, imino groups present are to some extent converted to chloramine groups:

$$>\!NH + HOCl \rightarrow\, >\!NCl + H_2O$$

These are unstable and break down on ironing, giving free hydrogen chloride which tenders the fabric. When a melamine resin is present

AMINO RESINS

chloramines are again formed and, although these are more heat stable and do not break down on ironing, they are bright yellow in colour.

A general solution to what is commonly referred to as the 'chlorine retention' problem has been sought in examining the methylol derivatives of amino compounds which contain no imino groups. Compounds that have been used commercially are shown in Fig. 2.4.

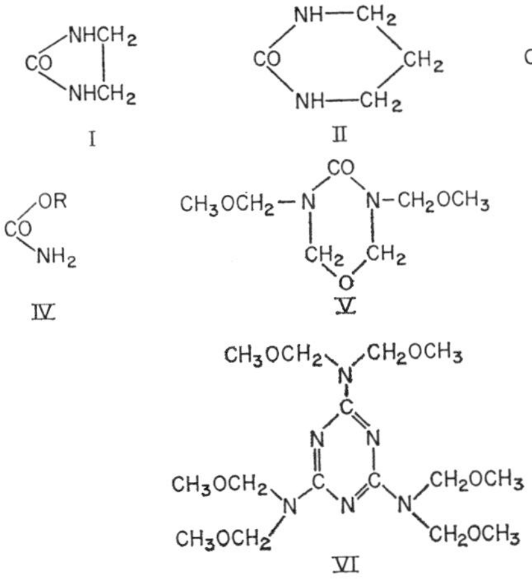

FIG. 2.4.

Ethylene urea (I), made by reacting urea with ethylene diamine, readily reacts with formaldehyde, yielding a dimethylol derivative which has a melting point of 133°C and dissolves readily in water. Good crease resistance and smooth-drying finishes are obtained using preferably a metal salt, such as magnesium chloride or zinc nitrate, as a catalyst. There is now ample evidence to suggest that dimethylol ethylene urea reacts with adjacent cellulose molecules forming cross-links. Initially, fabrics treated with this agent are non-chlorine retentive. However, after several washings the fabric becomes increasingly chlorine retentive. This has been shown to arise as a result of hydrolysis and breakage of the cross-links and not through a breakdown of the cyclic structure. If the treated fabric is washed in aqueous sodium carbonate after baking, a small amount of magnesium carbonate or zinc carbonate is formed by reaction with the metal salt catalyst. This tends to react with any hydrogen chloride formed on ironing and prevents weakening of the fabric. However, a further deficiency of ethylene urea is that it tends to impair the lightfastness of some direct and reactive dyes.

Propylene urea (II), made by reacting urea and 1,3-diamino propane, also reacts readily with formaldehyde and the dimethylol compound gives finishes which are more durable to repeated launderings than dimethylolethylene urea. However, it is considerably more expensive than the former and has the same disadvantage of impairing the lightfastness of dyes.

Dihydroxyethylene urea (III) is made by reacting glyoxal and urea, though it is customary to form its dimethylol derivative by coreacting urea, glyoxal and formaldehyde in 1:1:2 molar ratio without isolating (III). This is probably the most widely used of the cyclic urea finishes. Cross-linking with cellulose occurs not only via the methylol groups but also to some extent via the ring hydroxyls. Little formaldehyde odour is noticed with this product during finishing and after curing, and the lightfastness of direct and reactive dyestuffs is not impaired. Dimethylol dihydroxy ethylene urea finds its greatest outlet in the production of durable press garments, which will be mentioned later. Zinc nitrate appears to be the preferred catalyst.

Methyl and ethyl carbamates (IV, $R = CH_3$ or C_2H_5) react with excess formaldehyde, giving dimethylol derivatives. These, applied to cellulose fabrics using a metal salt catalyst, give very good smooth-drying finishes having excellent resistance to chlorine damage and to hydrolysis on laundering. There is evidence, however, that ethyl carbamate and its methylol derivatives are carcinogenic[10] and though this would not necessarily constitute a danger when they are reacted with cellulose, the dimethylol derivative of methyl carbamate, which is not carcinogenic, is clearly preferred.

N, N'-bis(methoxy methyl)uron (V) was first isolated by Kadowaki in 1936.[11] It is normally encountered as a colourless liquid with a boiling point of 82°C at 0·1 mm. When carefully purified it forms a solid melting at 29°C. It is prepared by reacting urea and formaldehyde (as formalin) in 1:4 molar ratio under alkaline conditions, removing water by distillation and treating the residue in the cold with a large volume of methanol containing hydrochloric acid. It is a difunctional compound with no free NH groups and gives good levels of crease resistance and is reasonably wash-fast. However, it is rather troublesome to prepare and is somewhat expensive. The structures of (V) and the parent uron (VA), which is obtained from (V) by acid catalysed hydrolysis using dimedone,[12] have been confirmed by NMR.[13]

The hexamethyl ether of hexamethylol melamine (VI), which is discussed more fully later in this paper, has no free imino groups, and has been used, blended with dimethylol ethylene urea, in commercial non-chlorine retentive textile finishes.

An interesting development in textile finishing in recent years has been the introduction of durable press garments. The earliest of several methods employed for manufacturing durable press clothing is the Koratron process, introduced commercially in the USA in 1964 by the Koret company of California. Briefly this consists of impregnating fabrics, made from blends of cotton or rayon with polyester fibre, with a solution of a cross-linking amino resin and a metal salt catalyst and then drying carefully to avoid substantial resin fixation. The fabric is made into garments which are then hot pressed and finally baked for 10 to 20 min in an oven at 145–155°C. Quite clearly, in a process of this kind the amino constituent must not hydrolyse or cross-link while the impregnated and dried fabric is awaiting fabrication into the required garments; nor can the finished garments be given a final wash after baking.

The dimethylol derivative of dihydroxyethylene urea (III) has been found to be particularly suitable for durable press processes, since it is able to exist in an unreacted state in the presence of a metal salt catalyst for long periods.

2.4.3 Coating resins

Butylated amino resins have been used for many years as film-forming components of pigmented stoving enamels employed as finishes for cars, refrigerators and many other commodities. A butylated resin alone, on stoving, gives a very hard, colourless, brittle film with poor adhesion to a metal substrate. Used in conjunction with an oil-modified alkyd resin or a thermosetting acrylic resin with pendant hydroxyl groups, the film obtained is still hard but is flexible and has good adhesion to metal. It is probable that in stoving (usually at temperatures in the range 120–145°C) the butylated resin both undergoes self-condensation and coreaction with the alkyd or acrylic resin via their free hydroxyl groups.

Butylated resins may be made by various techniques but the method most often used is as follows. Urea or melamine is first dissolved in neutralised formalin, excess butanol and a small amount of an entraining agent (usually xylene or white spirit) are introduced,

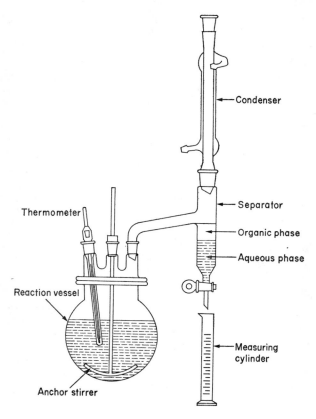

Fig. 2.5. The preparation of butylated amino resins.

and water is removed azeotropically using a decanter system similar to that shown in Fig. 2.5.

The azeotrope which distils over at a temperature of c. 92°C (i.e. well below the boiling point of n-butanol) condenses and separates into two layers. The upper layer containing butanol, xylene and a little water is returned to the reaction vessel; the lower layer consisting of

AMINO RESINS

water and a little dissolved butanol is withdrawn as necessary. When most of the water has been removed, an acid catalyst (e.g. a solution of phthalic anhydride in butanol) is added and azeotropic distillation continued. Butylation of methylol groups occurs concurrently with various condensation reactions to produce eventually a clear colourless resin of required organic solvent tolerance, viscosity and curing rate.

The chemical nature of the products obtained is illustrated by data given in Table 2.5. In an experiment using substantially the technique

TABLE 2.5

Sample	Reaction time (h)	Tolerance (ml white spirit per 5 g resin)	Nitrogen content (%) of dry resin (N)	Number average mol. wt. (M)	Degree of polymerisation (n)	No. of substituents per melamine nucleus	
						CH$_2$O	OBu
1	1	5	26·1	1350	4·2	3·9	1·6
2	2	10	25·6	1340	4·1	4·1	1·7
3	4	20	24·4	1250	3·6	4·2	1·9
4	8	32	23·2	1340	3·7	4·3	2·3
5	16	83	22·4	1890	5·1	4·3	2·5
6	26	>150	21·4	2530	6·5	4·5	2·8

described above, 1 mol. melamine was reacted with 5.3 mol. formaldehyde (as formalin) and 5.4 mol. n-butanol. Samples were withdrawn at different times after the addition of the acid catalyst. The tolerance for white spirit was measured and the sample dried under vacuum at room temperature to constant weight. The product was then analysed for nitrogen content (Kjeldahl), total and free formaldehyde, and butoxy group content (Zeisel), and the number average molecular weight determined using a vapour pressure osmometer. From the various results obtained the figures given in Table 2.5 have been calculated. The degree of polymerisation, n, is the average number of triazine nuclei present per molecule. Since there are $6n$ nitrogen atoms present per resin molecule, the percentage nitrogen present, N, is given by

$$N = \left(\frac{6n \times 14}{M}\right) \times 100$$

where M is the number average molecular weight. Thus

$$n = \frac{MN}{8400}$$

It will be observed that the amount of butylation increases with increasing reaction time up to 8 h or so, with no apparent growth in molecular size. Thereafter, condensation reactions appear to predominate and a substantial increase in molecular size and in viscosity occurs. In general, the greater the degree of butylation the better the compatibility with white spirit but the slower the rate of cure at a given stoving temperature and, in practice, a compromise has to be reached between these opposing factors.

n-Butanol (boiling point 117°C) and *iso*butanol (boiling point 108°C) are both used in the preparation of commercial resins, the latter being less expensive, and resins obtained tending to cure more rapidly, possibly because of the greater volatility of the alcohol. Secondary butanol is far less reactive, tending to give highly viscous resins with inferior solvent compatibility,[14] and is not used commercially. Resins based on higher alcohol homologues such as 2-ethyl hexanol are made by reacting the butylated resins with the higher boiling alcohol in the presence of a transetherification catalyst, then distilling off the liberated butanol. However, resins of this kind have achieved no commercial importance.

Alkylation with the lower alcohols, particularly methanol, gives in general water-soluble resins which have been used widely in textile finishing but are not compatible with hydrocarbon solvents.

A product, however, which has aroused much interest in recent years is the hexamethyl ether of hexamethylol melamine (HMEHMM), having the structure (VI) in Fig. 2.4. It can be made in a pure state by dissolving melamine (1 mol.) in excess formalin (9 mol. CH_2O) at about 60°C, the pH being maintained at 9·0–9·5. The hexamethylol melamine, which separates on cooling, is filtered, washed with water, and oven-dried at temperatures not exceeding 40°C. The dry product is treated with a considerable excess of methanol under acid conditions until dissolution is complete, the solution then being neutralised and filtered, and excess methanol removed by vacuum distillation. The pure ether is a colourless, crystalline solid melting at 55°C. Commercial grades, available as waxy solids or colourless liquids, are to some extent condensed and contain a small proportion of unreacted amino hydrogens. An analysis of one such commercial product showed it to contain 5.8 mol. combined formaldehyde per mol. melamine, and 5.25 mol. methoxy methyl groups.

HMEHMM has only a limited solubility in water but dissolves readily in water containing methanol. It is also soluble in a wide range of organic solvents. Heated with a strong acid catalyst it yields an infusible thermoset product with liberation of methanol. It also reacts, in the presence of acid catalysts, with compounds containing a hydroxyl group, transetherification occurring with loss of methanol.

The ability to react with groups containing a reactive hydrogen atom, together with its wide range of solubility and polymer compatibility, enables it to be used as a cross-linking agent with a number of different polymers in coating formulations. However, to obtain practical cure cycles it is necessary to incorporate a strong acid catalyst such as *p*-toluene sulphonic acid. Even with commercial products most of the functional groups of HMEHMM are used for cross-linking and not self-condensation and, for example, in conjunction with alkyds cured films have greatly improved flexibility. HMEHMM, unlike most commercially available MF resins, is freely compatible with commercial epoxy resins, reaction with hydroxyl groups, and not with epoxide groups, occurring on curing. Hard films with good flexibility and resistance to alkalis are obtained using these blends, the optimum amount of HMEHMM being about 15% of the resin mixture.

HMEHMM has also been widely recommended as a cross-linking agent for water-soluble alkyds or water-soluble acrylics used in the production of water-based coating resins.

REFERENCES

1. Business Monitor; Production Series, 'Synthetic Resins and Plastics', H.M.S.O., 1972–73.
2. Private communication, B.I.P. Market Research and Development Group.
3. Stern, J. P. and Stern, E. S. (1971). *Petrochemicals Today*, Arnold, London.
4. Slack, A. V. and Blouin, G. M. (1971). *Chem. Technology*, January, 32.
5. Biddle, J. W., C.I.O.S. XXXI-81, H.M.S.O., 1945.
6. Baumann, H. (1957). *Kunststoffe*, **47**, 256; (1958) **48**, 362, 406.
7. *Europlastics Monthly* (1974). **47** (1), 50.

8. *Resin News* (1973). **13** (10), 6.
9. Smith, A. R. (1967). Private communication and *Rep. Progress of Appl. Chem.*, **52**, 663; (1969) **54**, 560; (1971). **56**, 231.
10. Booth, A. N., *et al.* (1967). *Amer. Dyestuff Reptr.*, **56**, 73.
11. Kadowaki, H. (1936). *Bull. chem. Soc. Japan*, **11**, 248.
12. Beachem, M. T., *et al.* (1963). *J. Org. Chem.*, **28**, 1876.
13. Egginton, C. D. and Vale, C. P. (1969). *Text Res. J.*, **39**, 140.
14. Seaborne, L. R. (1955). *J. Oil Col. Chem. Ass.*, **38**, 345.

CHAPTER 3

Recent Developments in Unsaturated Polyester Resins

K. Parvin

(Scott Bader Co. Ltd)

3.1 INTRODUCTION

The commercial importance of unsaturated polyester resins has steadily increased over the years and UK production in 1973 was in excess of 50 000 tonnes. Although the basic formulation of general-purpose polyesters has changed little during the years, our knowledge of the effect of constituents on properties and the cross-linking reaction has steadily improved.

It would be impossible to cover the whole field and three items have been selected which are of current interest: (*a*) catalyst/accelerator/inhibitor systems; (*b*) low-profile polyester moulding compositions; (*c*) resins of reduced fire hazard.

3.2 CATALYST/ACCELERATOR/INHIBITOR SYSTEMS

The terms used in the industry have been retained, although the correct title of this section should be initiator/promoter/inhibitor systems. Until recently, catalyst/accelerator systems have been considered separately from the inhibitors used but they cannot really be separated as can be seen from Table 3.1.

Of the three common inhibitors benzoquinone behaves quite differently to the other two when cured with benzoyl peroxide at 82°C, whereas at room temperature with methyl ethyl ketone peroxide (MEKP) and a cobalt soap its behaviour is similar to that of *t*-butyl catechol.

Again, the behaviour of a compound can change from that of a promoter and become an inhibitor at a different concentration or temperature. This is illustrated in Table 3.2, where it can be seen that a quaternary compound in low concentrations acts as promoter

TABLE 3.1

EFFECT OF INHIBITORS

Inhibitor (100 ppm)	Benzoyl peroxide 82°C			MEKP/cobalt 25°C		
	G.T. (min)	C.T. (min)	Peak temp. (°C)	G.T. (min)	C.T. (min)	Peak temp. (°C)
—	2·0	4·0	205	3·4	10·4	132
Hydroquinone	2·75	4·9	205	3·25	10·75	168
Benzoquinone	5·0	7·4	207	7·75	17·75	154
t-Butyl catechol	2·1	4·1	202	7·0	16·25	131

G.T. = gel time. C.T. = cure time, taken as the time when the exotherm reaches its highest value.

TABLE 3.2

EFFECT OF A QUATERNARY COMPOUND IN CONJUNCTION WITH HYDRO-QUINONE IN A ROOM TEMPERATURE CURING SYSTEM

BTMAC (ppm)	Gel time (min)	Cure time (min)	Peak exotherm temp. (°C)
—	36·3	57·9	110
200	33·3	49·8	118
600	31·7	49·4	121
1000	42·0	56·6	119
1400	53·8	81·5	103
1600	86·8	134·0	63

BTMAC = benzyl trimethyl ammonium chloride.

and at higher concentrations as an inhibitor (copper compounds can behave in the same manner).

It is not proposed to continue discussing inhibitor systems, as these are usually chosen by the resin manufacturer and incorporated at the blending stage. The moulder does have some control on the promoter system used and, in recent years, dual promoter systems have become of interest for fast hardening resins to achieve faster mould turn-round. What is not well appreciated is the effect of amine/cobalt ratios on the gelling and hardening characteristics of a resin. Table 3.3 shows the effect of various amine and cobalt levels on the curing characteristics of a polyester resin.

From experiments 1–3 it can be seen that, at a given amine level, cobalt additions lengthen the cure time and the laminates do not harden so rapidly. At lower amine levels the cure times shorten instead of lengthening (see experiments 4–6, 7–9). If the same cobalt/

TABLE 3.3
EFFECT OF AMINE/COBALT RATIOS

Expt. no.	Cobalt octoate (%)	Diethyl aniline (%)	Gel time (min)	Cure time (min)	Peak temp. (°C)	Laminate hardness (Barcol)	
						2 h	4 h
1	0·075	0·10	9·0	7·5	96	32	37
2	0·150	0·10	10·2	18·0	86	27	33
3	0·300	0·10	11·4	20·1	70	23	30
4	0·075	0·05	10·0	27·3	76	15	24
5	0·150	0·05	11·5	24·5	63	16	24
6	0·300	0·05	10·0	22·5	56	4	12
7	0·075	0·025	12·0	33·0	63	16	21
8	0·150	0·025	9·0	25·0	57	11	21
9	0·300	0·025	9·5	24·0	50	65[a]	76[a]

[a] Shore D.

amine ratio is maintained, differences in gel time and hardening rate are not so apparent (see experiments 3, 5 and 7).

When considering initiators a wide range of gel and cure times can be obtained on the same resin by the use of different peroxides, as can be seen from Fig. 3.1.

3.3 LOW-PROFILE RESIN SYSTEMS

The terms low-profile and low-shrinkage systems are often used interchangeably to describe certain types of moulding compositions, such as sheet mouldings compounds (SMC) and bulk moulding compounds (BMC), which yield mouldings with excellent surface finish.

Shrinkage on polymerisation occurs with all addition-type polymerisations due to the fact that in the monomers the molecules are located at a van der Waals' distance from one another, while in the polymer the units move to within a covalent distance of each other, i.e. the atoms are much closer to one another in the polymer than in the original monomer. By comparing the specific gravity of

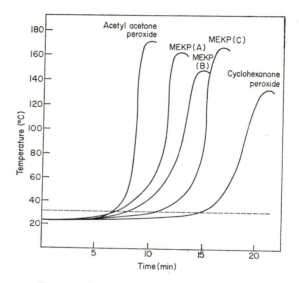

Fig. 3.1. Exotherm curves for Crystic 471 PA.

monomers and polymers the shrinkage on addition polymerisation can be calculated, and some figures are given in Table 3.4.

It can be seen from Table 3.4 that, in general, as the molecular weight is increased the shrinkage decreases. In an unsaturated polyester resin the amount and type of monomer employed controls the shrinkage of the resultant solution in monomer and a diagram can be constructed showing the expected shrinkage of a polyester with different monomers and amounts of monomers (see Fig. 3.2).

It can be seen that for a standard styrenated polyester with 35% monomer the shrinkage should be 8%, the usually accepted figure.

TABLE 3.4
CALCULATED SHRINKAGES FOR HOMOPOLYMERISATION

Monomer	Volumetric shrinkage (%)	Molecular weight
Ethylene	66	28
Vinyl chloride	34·5	62·5
Vinyl acetate	27	86
Acrylonitrile	26	53
Methyl methacrylate	21	100
Styrene	17	104
Monochlorostyrene	13	138·5
Vinyl toluene	12·6	118
Diallyl phthalate	11·8	246
t-Butyl styrene	7	160
Unsaturated polyester	3	2000

This can be reduced to 6·4% by using vinyl toluene or 4·2% using t-butyl styrene. However, because the two latter monomers yield polyesters with higher viscosities than styrene-based resins, higher monomer contents will be required to maintain a given viscosity and this will increase the shrinkage above the figures mentioned. Another complicating factor is that not all polyesters are soluble in t-butyl styrene.

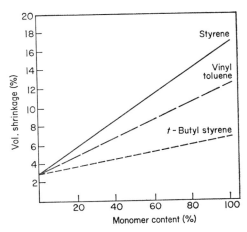

Fig. 3.2. Dependance of volumetric shrinkage upon monomer content.

The shrinkage of moulding compositions such as SMC or BMC is also reduced by the fillers which are incorporated, usually at a weight equal to that of the resin used. This means that the shrinkage will be about half that of the unfilled resin.

Mouldings from standard grades of SMC and BMC have good surface finish but not the excellent finish that is required in certain applications such as automobile body components, furniture and consumer-durable housings. This is obtained by the incorporation of thermoplastic polymers into the compositions. From their inception SMC and BMC formulations have incorporated small quantities, $c.$ 3%, of polyethylene but of recent years the uses of numerous other thermoplastics have been the subject of patents, usually with the object of obtaining a low surface profile.

The first claim of this kind was made in a French patent, dated 1957, which covered the use of thermoplastic polymers soluble in, or swollen by, styrene, to reduce shrinkage, and the examples given included polystyrene, polyvinyl ether, polyvinyl carbazole and polyisobutylene. Since that date practically every thermoplastic has been covered in the patent literature and a few examples of materials currently used will be considered. Polyvinyl acetate homopolymers and copolymers are being offered as low-profile additives by Union Carbide. Similar materials have formed the subject of patent claims by Diamond Shamrock.

Saturated polyesters are the subject of patents to both Koppers and a Japanese company, whilst somewhat different polyesters, namely polycaprolactones, are marketed by Union Carbide as low-profile additives.

Rohm and Haas, in a series of patents all over the world, have claimed the incorporation of a thermoplastic, soluble in styrene, which forms an optically heterogeneous, two-component system on curing. The examples given include acrylate copolymers and cellulose acetate butyrate. However, Eastman Kodak have promoted some grades of cellulose acetate butyrate as low-profile additives.

From this brief survey it can be seen that the patent position is extremely complex with a wide range of thermoplastics covered in claims. However, it is known that some additives suffer disadvantages such as giving surfaces with poor paint adhesion or systems that are not suitable for pigmentation. To date, no one has published a direct comparison of the various additives, which could be very interesting.

3.4 RESINS OF REDUCED FIRE HAZARD

This term is used deliberately as it is now the designation approved by British Standards and supersedes such terms as fire retardant, self-extinguishing, etc. which are considered misleading.

It has been known for many years that the fire hazard of polyester resins can be reduced by the incorporation of halogens together with synergists such as antimony oxide. The commonest and cheapest way of obtaining such resins has been by the use of chlorinated paraffin and antimony oxide or, if clear resins were required, by the addition of halogenated phosphates such as trichloroethyl phosphate. Such resins would, when reinforced with glass fibres, give laminates which conformed to Class II of BS 476 Part 7.

To obtain improved properties chlorine was then incorporated into the polyester chains by the use of tetrachlorophthalic anhydride or chlorendic acid but these resins, when combined with synergists, still only yielded laminates obtaining Class II ratings. To enable polyester laminates to be used in buildings it was necessary to develop resins which would obtain a Class I rating, thus complying with the Building Regulations. For this reason numerous other means of introducing halogens, either by means of additives such as pentabromotoluene and tris(dibromopropyl)phosphate or by reactants such as tetrabromophthalic anhydride or dibromoneopentyl glycol, were investigated. The effects of alternative synergists and inert fillers were also examined, with the result that both clear and opaque resins are now available which can be made into laminates capable of obtaining a Class I rating.

One of the problems encountered in these developments was finding a suitable laboratory scale test for differentiating between the burning properties of the various formulations. The BS 476 Surface Spread of Flame Test calls for six specimens, each 900 mm × 230 mm; the manufacture and testing of these laminates at one of the testing stations is quite expensive and formulation changes cannot be checked in this manner. It also became obvious that simple burning tests, e.g. the strip burning test described in Appendix B of BS 3532, could not differentiate between formulations with very low fire hazard. For this reason the Limiting Oxygen Index test was adopted.

This test consists of burning a vertical sample in a candle-like manner in a mixture of oxygen and nitrogen. The volume percentage of oxygen in the mixture at which the specimen just continues to

Fig. 3.3. Apparatus used for determining the Limiting Oxygen Index.

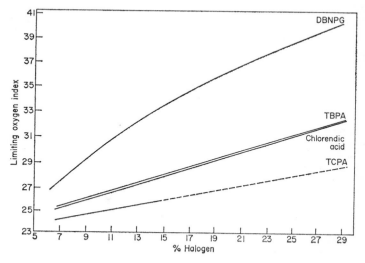

Fig. 3.4. Halogen content against Limiting Oxygen Index (all resins contain 1% P as TEP). DBNPG—dibromoneopentyl glycol, TBPA—tetrabromophthalic anhydride, TCPA—tetrachlorophthalic anhydride.

burn is taken as the Limiting Oxygen Index (LOI). Using this test the differences between various systems of reducing the fire hazard become readily apparent (Fig. 3.3). Figure 3.4 shows plots of oxygen index against halogen content for a number of different reactive components compounded with triethyl phosphate to give 1% phosphorus in the resin. It can be seen that there is little difference between chlorine incorporated as chlorendic acid or bromine incorporated as tetrabromophthalic anhydride, whereas dibro-

TABLE 3.5

OXYGEN INDEX OF DIFFERENT SYSTEMS

Halogen source	Halogen content (%)	LOI 1% P	LOI 5% Sb_2O_3
Chlorinated paraffin	7·5	23·8	24·2
	15	—	27·4
Chlorendic acid	7·5	23·1	27·2
	15	27·2	30·2
TCPA	7·5	21·7	23·0
	15	25·0	26·1
DBNPG	7·5	24·9	25·8
	15	33·0	30·6
TBPA	7·5	22·8	25·8
	15	27·5	29·7
Pentabromotoluene	7·5	22·7	25·7
	15	25·8	31·0

moneopentyl glycol is much superior. This is not the case with other synergists and there are some published data showing that bromine is more efficient than chlorine with certain inorganic materials such as antimony oxide.

We have recently done some work on the incorporation of halogens at fixed levels by various methods and using different synergists. Some of the results are shown in Table 3.5.

It is too early to arrive at any conclusions but it is evident that differences occur and these are undoubtedly due to differences in the breakdown mechanisms of the various systems.

Whilst the reduction of fire hazard is important, consideration is also being given to smoke generation and toxic fumes. In general, the incorporation of fire retardants increases the smoke generated from burning resins. Some inorganic additives such as alumina hydrate delay the production of smoke, i.e. they delay the time to maximum obscuration in tests to ASTM D2643. Further developments on these systems are expected.

Finally, an alternative method of fire protection is now being used, employing intumescent coatings. These can be based on polyester resins with three additives: (*a*) a source of phosphoric acid; (*b*) a polyhydroxy compound which reacts with phosphoric acid to form a char; and (*c*) a blowing agent. On the application of a flame these coatings yield a carbonaceous expanded char which protects the underlying laminate, and it is easy to obtain structures with Class I or Class 0 ratings.

CHAPTER 4

Friedel–Crafts Resins

A. G. Edwards
(*Albright and Wilson Ltd*)

4.1 INTRODUCTION AND DEVELOPMENT

The classical Friedel–Crafts reaction involves the introduction of an alkyl or acyl group into an aromatic ring in the presence of certain catalysts. A simple example is the reaction of benzene with methyl chloride to give toluene (Fig. 4.1).

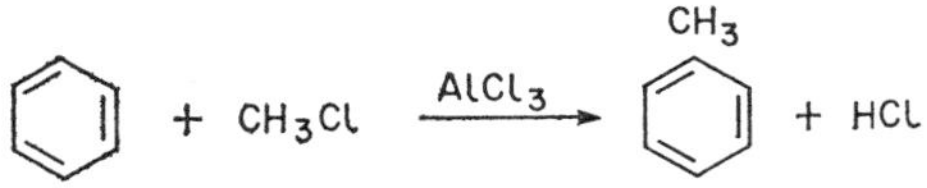

Fig. 4.1.

Aluminium chloride is the most commonly used catalyst but it is only one of a number of compounds that will bring about this type of condensation.

The application of this reaction to make a resin stems from a discovery made by Friedel and Crafts as long ago as 1885.[1] They heated benzyl chloride in the presence of aluminium chloride and obtained an insoluble infusible mass (Fig. 4.2).

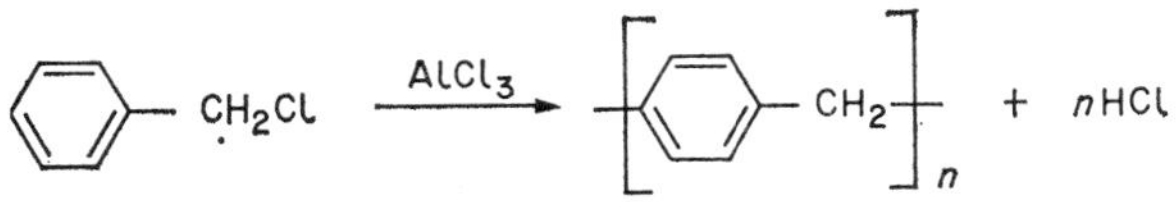

Fig. 4.2.

Benzyl chloride has only one chloromethyl group and it is clear that the polybenzyl shown in Fig. 4.2 cannot be a true thermosetting material, since cross-linking cannot take place. In fact the structure was later proved to be a very highly branched molecule with some of the aromatic rings substituted several times.[2] This structure gives polybenzyl some interesting properties; for example, weight loss studies have shown that it is more thermally stable than phenolic resins[3] but the lack of cross-links leads to poor mechanical strengths at elevated temperatures. This last-mentioned fact, coupled with

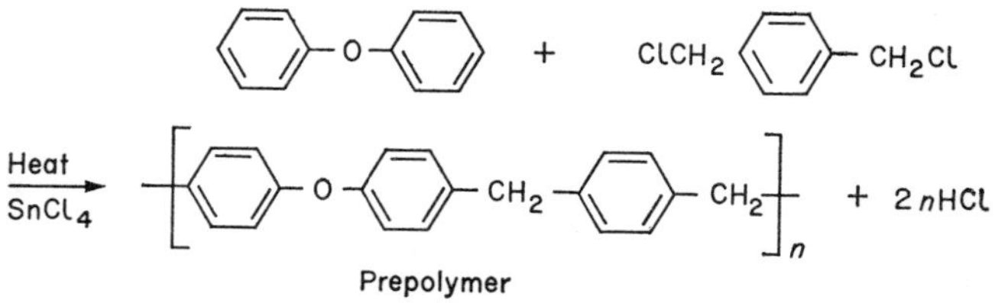

FIG. 4.3.

the inherent processing difficulties caused by the evolution of hydrogen chloride during the preparation, meant that the polymer remained only of academic interest.

The discovery of the phenolic, amino, polyester, epoxide and silicone resins largely enabled the demands of technology to be met up to the late 1950s. At about this time, however, technological demands of aircraft and aerospace industries prompted a renewed search for strong, more heat resistant, plastics materials.. It was soon discovered that in order to prepare organic polymers that would have a useful life at temperatures above 200°C, it was necessary to have a structure based on a large number of aromatic and/or heterocyclic rings. Thus, interest in the Friedel–Crafts system was revived by a number of companies and organisations both in the USA and in the UK.[4–7]

This time the concept was extended in order to produce true thermosets by using compounds having two or three chloromethyl groups. Other aromatic compounds were also included in the polymer structure. An example is shown in Fig. 4.3, where diphenyl ether is condensed with $\alpha\alpha'$-dichloro-p-xylene in the presence of stannic chloride.

A prepolymer is formed initially and this is cross-linked by the addition of more αα'-dichloro-*p*-xylene. The problem of hydrogen chloride evolution during the final cure was still present and this was obviously not going to be acceptable commercially. The search for alternative reactive intermediates led to the simultaneous, but independent, discovery of the aralkyl ethers, in 1964, by chemists at the Royal Aircraft Establishment, Farnborough, and Albright and Wilson Ltd. These new materials were found to react in a similar manner to the chloromethyl compounds but the by-products of the reactions are alcohols.[8,9]

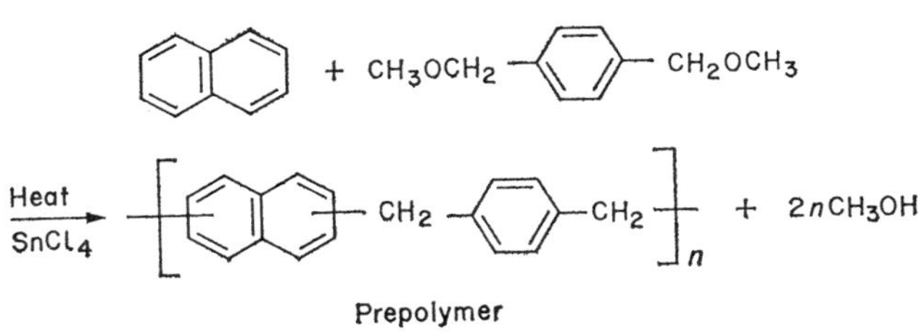

$$\text{naphthalene} + CH_3OCH_2\text{—}\langle\text{C}_6H_4\rangle\text{—}CH_2OCH_3$$

$$\xrightarrow[SnCl_4]{\text{Heat}} \left[\text{—naphthylene—}CH_2\text{—}\langle\text{C}_6H_4\rangle\text{—}CH_2\text{—} \right]_n + 2n\,CH_3OH$$

Prepolymer

FIG. 4.4.

This is illustrated by the reaction of naphthalene with αα'-dimethoxy-*p*-xylene, where methanol is eliminated (Fig. 4.4).

Again a prepolymer is formed and cross-linking is effected by the addition of further aralkyl ether and heating in the presence of a stronger Friedel–Crafts catalyst, such as ferric chloride. Further amounts of the alcohol are eliminated during the final cure. As the main processing difficulties had now been overcome resins of this type have been used commercially.[10,11] Applications have tended to be limited to those requiring thermal stability rather than high temperature strength, e.g. insulating varnishes for electric motors. This is because the curing rate is relatively slow and the cross-link density is probably inadequate.

During a recent study of the use of Friedel–Crafts resins for the preparation of carbon fibre reinforced composites, attempts were made to overcome these disadvantages. Some improvement in the rate at which the high temperature strength develops was obtained by using resins based on mononuclear aromatic compounds, such as toluene, and increasing the curing catalyst concentration.[12,13]

However, the real breakthrough in the exploitation of Friedel–Crafts resins was achieved when Albright and Wilson Ltd decided to concentrate on the final cross-linking reaction. By condensing the aralkyl ethers with phenolic compounds a prepolymer is formed that can be cross-linked in a number of ways. One of the simplest examples is shown in Fig. 4.5, where the prepolymer is formed by condensing $\alpha\alpha'$-dimethoxy-p-xylene with a molar excess of phenol.

The structure of the prepolymer can be varied widely by the use of different phenols, mixtures of phenols or even combinations of phenols with other classes of aromatic heterocyclics or organometallic compounds.

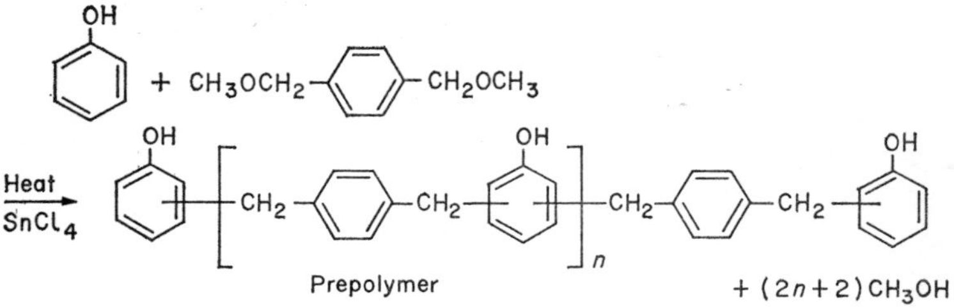

FIG. 4.5.

Although these resins can be cured by several different techniques, only two methods are currently being used commercially. In the first method the resin is heated with hexamethylene tetramine (hexamine) which attacks the vacant *ortho* and *para* positions on the phenolic nuclei with the formation of dimethylene amino linkages.[14] This is accompanied by the release of ammonia. Slow decomposition of these linkages at elevated temperatures results in the formation of the more stable methylene and azomethine groups and more ammonia. The second method of obtaining a cross-linked product is by heating with selected epoxide compounds in the presence of an accelerator.[15] Reaction occurs between the epoxide groups and the phenolic hydroxyls without the formation of any by-products.

The products obtained by the different curing mechanisms† will be considered separately.

†Sold under the registered trademark, Xylok resins.

4.2 HEXAMINE-CURED RESINS

The resins of this type are mainly used for the preparation of reinforced laminates or high-performance moulding compounds. Since the curing mechanism is similar to that used with phenolic resins then they share the same easy processing characteristics and rapid development of high temperature strength. Contrary to expectation, however, this is achieved without any significant loss of the thermal stability of the other Friedel–Crafts resins.

4.2.1 Reinforced laminates

Preparation and properties

Laminates can be made using all forms of glass and asbestos reinforcements. The detailed preparation and properties of these have been given before and are well known.[16] Thus only a brief outline will be presented here.

The main characteristics of the composites are listed below:

(*a*) excellent thermal stability—suitable for prolonged exposure at temperature of 150–250°C;

(*b*) good mechanical strength and high-temperature rigidity;

(*c*) low, stable values for permittivity and loss tangent;

(*d*) outstanding machining and punching characteristics;

(*e*) low water absorption and excellent dimensional stability;

(*f*) good resistance to attack by a range of chemical reagents, hydraulic fluids and engine lubricants;

(*g*) high radiation resistance.

All of these properties can be explained in terms of the high aromatic content of the resin structure and/or the rapid cross-linking reaction. Obviously the most interesting properties are the mechanical strength and strength retention at high temperatures. These are best illustrated in comparison with results obtained from boards based on other commercially available thermosets, as shown in Fig. 4.6.

The flexural strengths of these glass-reinforced laminates were measured at 250°C after heat ageing at this temperature. The results show that the materials fall into three classes. In the first class are the phenolic and special high temperature phenolic resins which have high initial strengths but poor thermal stability. The second class comprises the epoxide, epoxy novolac, silicone and thermosetting acrylic which have relatively poor strengths but the strength retention is good. Only two resins, the Friedel–Crafts and polyimide, come into the last class, where both the mechanical strength and the strength retention are high. Over 80% of the initial strength is retained by the Friedel–Crafts resin after 1000 h at 250°C and even after 2000 h at this temperature over 50% of the strength remains. At even higher temperatures the Friedel–Crafts boards still have an impressive performance. It takes approximately 800 h at 275°C and 300 h at 300°C for the initial high temperature strength to be reduced by 50%.

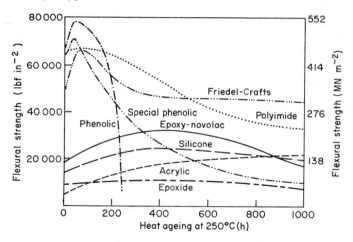

Fig. 4.6. Flexural strengths (measured at 250°C) of glasscloth laminates heat aged at 250°C.

Applications

In terms of cost these Friedel–Crafts resin composites fall roughly between those of epoxides and silicones. When it comes to actual applications, it is rarely any one property of the laminates that is of importance but, more likely, a combination of several of the factors presented in the previous section. The electrical industry in particular provides many examples where good electrical insulation properties are required in conjunction with high strengths and thermal stability. Thus, it is not surprising that boards made from the hexamine-cured Friedel–Crafts resins have provided many answers to insulation problems.

Glasscloth-reinforced laminates are being used as slot wedges in electrical motors because (a) their high strengths permit them to be hammered into position, (b) they can withstand the centrifugal forces and (c) they have a long life under the operating conditions. Tubes made from the same material are used in the manufacture of gas reservoirs in high voltage switchgear. An inert gas is stored in the tubes at a pressure of about 1000 lbf in^{-2} (6·89 MN m^{-2}) and this is released between the conductors to prevent arcing during the switching off operation. Temperatures as high as 300°C are reached for a short time.

The Friedel–Crafts resins cured with hexamine can be flexibilised by the addition of a small quantity of an epoxide resin. This fact is used when insulating traction motors for the new British Rail 25 kV locomotives, type AL7. The resin is used as a varnish on the copper windings and the insulation is completed by using asbestos and glass tapes impregnated with the resin. The motors are rated at 1250 HP, 885 A and 1134 V and are designed to Class H insulation standards, i.e. capable of withstanding continuous operation at 180°C.

A whole new range of transformers has been produced using laminates based on these resins. One of the transformers was designed for operation in the Concorde aircraft. Previous experiments have shown that these boards have outstanding resistance to transformer oil. There was virtually no change in the acid value of the oil even after 12 000 h at 200°C.

An important application area where the excellent thermal stability, electrical properties and high temperature strength of these composites are being exploited is in industrial and domestic heaters. Metal foil or wire elements are electrically insulated by sandwiching them between high resin content glasscloth prepregs. These heaters are being extensively used in textile equipment as are other parts machined from glasscloth laminates. These latter applications require a long life at 250°C and good abrasion resistance.

The electrical industry is not the only outlet for these resins. Severe operating conditions have led to a number of applications in engines. These have usually been in disel engines and involve resistance to various warm diesel fuels and lubricants.

Friction linings have been made from phenolic resins for many years but the demand for constant frictional properties at high temperatures has led to the use of the hexamine-cured Friedel–Crafts resins in the formulation of brake-lining materials. The improvement in performance obtained is at least partly explained by the fact that a higher filler content can be used. The disc brakes of a well-known European car are based on this material and they exhibit improved 'antifade' characteristics.

4.2.2. Moulding compounds

Preparation and properties

High performance moulding compounds made from the hexamine-cured resin are being produced commercially in Britain, Europe and Japan. A wide range of fillers and additives can be used to obtain materials suitable for many applications.

The fact that hexamine is used as the cross-linking agent is of even greater importance for moulding compounds than it is for laminates. Not only can they be prepared on conventional equipment as used for phenolics but, unlike many other new heat stable thermosetting moulding powders, they can be processed by compression, transfer and injection moulding techniques, using essentially the same conditions employed for phenolics.

In comparison with mouldings produced from similar phenolic moulding powders, those made from the Friedel–Crafts based compounds have similar mechanical strengths but much better thermal and dimensional stabilities, as well as superior dielectric properties. Details of these have been given before, so only the most important of them will be discussed here.[16-18]

As might be expected, the mechanical strength and thermal stability of the mouldings are significantly affected by the nature of the filler used. This is clearly shown in Fig. 4.7, where it can be seen that the highest strength is given by the use of chopped glass reinforcement, while silica-filled mouldings have the highest strength retention. The use of mica results in lower strengths but these mouldings have the best electrical insulation properties. Asbestos is the most versatile filler and is widely used in heat resistant applications.

Recently, work has been carried out on the use of blends of phenolic and Friedel–Crafts resins in the preparation of asbestos-filled moulding compounds. Since the resins have similar chemical structures and share the same curing mechanism there are no compatibility problems or processing difficulties. The effect of various ratios on the flexural strength retention on heat ageing at 250°C is shown in Fig. 4.8.

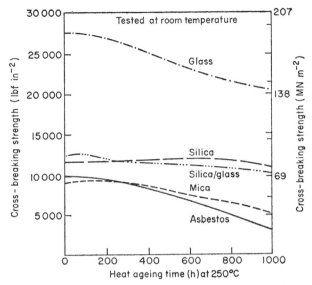

FIG. 4.7. Effect of filler on strength and thermal stability.

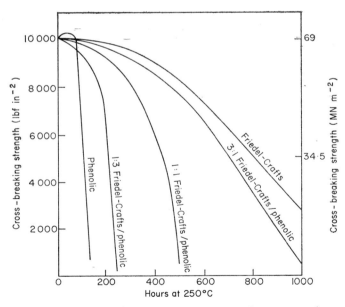

FIG. 4.8. Thermal stability of asbestos-filled moulding compounds.

The Friedel–Crafts compound gives mouldings which retain their strength at 250°C approximately eight times longer than that of a similar phenolic material. Replacement of 25% of the phenolic resin by the Friedel–Crafts polymer results in a 100% improvement in the thermal stability at 250°C. When equal quantities of the resins are used then the life at this temperature is 500 h. However, the most surprising result is that as much as 25% of the Friedel–Crafts resin can be replaced by the phenolic without there being any serious loss of strength retention. Incorporation of the phenolic resin brings about a reduction in cost and also increases the initial high temperature rigidity.

Using these results, a new injection-grade moulding compound has been developed, which has the very important property of producing mouldings which do not require postcuring. The moulding conditions needed are essentially the same as those employed for phenolics. The same short cure times (usually less than 1 min) are obtained but the components produced can be immediately exposed to temperatures as high as 280°C with very little shrinkage or change in appearance. The strength, electrical properties and chemical resistance are also highly developed without the need for postcuring.

Work is now in progress on extending this use of resin mixtures to other moulding compounds based on different fillers and suitable for both compression and transfer moulding. The eventual outcome of this work will be a complete range of materials that can be used to produce mouldings that do not require postcuring.

Applications

Applications for mouldings produced from compounds based on the hexamine-cured Friedel–Crafts resins are quite varied, but as with the laminates electrical insulation is very important. The use of these materials allows a greater safety margin in overload conditions which result in high temperatures.

A particular example occurs in commutators where moulding compounds are used to bond the central shaft and outer copper segments. The commutators used in the starter motors of diesel lorries can easily reach 200°C if the engine is difficult to start. Under these conditions, when phenolic insulation is used, the copper segments frequently move and this results in further overheating and progressive degradation. The use of an asbestos-filled Friedel–Crafts compound has resulted in commutators with greatly improved lives and they have been approved by the Ministry of Defence for use in military vehicles.

Considerable interest has been shown in the use of silica-filled Friedel–Crafts moulding compounds for the electronic encapsulation of semiconductor devices. The advantages over epoxide and silicone compounds include higher mechanical strength and greater solvent resistance. Where the devices are sensitive to low levels of basic material, they are first over-coated with an epoxide resin.

A growing application area for Friedel–Crafts mouldings is in bearings for operation in hot water pumps. Graphite is the preferred filler because it contributes to the dimensional stability and provides low frictional and wear properties. The use of these bearings is expected to be extended to driers, automotive components and guides in baking and textile equipment.

Mouldings for domestic appliances made from the new injection moulding compound based on the resin mixture described earlier, are generating widespread interest. The fact that the appearance is largely unaffected by immediate exposure to high temperatures without the need for postcuring means that they can be used to produce high quality handles for cooking utensils such as saucepans and casseroles. Their high dimensional stability reduces the risk of cracking caused by heat cycling and they show a much greater resistance to steam, humidity and detergent solutions than the moulded handles in current use.

The same Friedel–Crafts compound has been approved for producing the handles of industrial irons. Unlike domestic irons, these are left on for long periods of time and because of the high temperatures reached, the phenolic handles previously used had very limited lives.

Other applications for the Friedel–Crafts mouldings include those requiring a combination of thermal and chemical resistance. Examples of this type are electrode holders and handles for portable arc-welding equipment (which are showered with sparks of molten metal when in use), automobile components exposed to engine oil at 150°C for long periods and various parts of wet-type copying machines.

4.3 EPOXIDE-CURED RESINS

The most recent advance in Friedel–Crafts polymers has been the development of a system that cures by an epoxy mechanism, and a resin of this type has just been released commercially. A prepolymer based on the condensation of a phenol with an aralkyl ether is cured by reaction with an epoxide compound in the presence of a suitable accelerator. The epoxy groups react with the phenolic hydroxyls to form cross-links without any volatile by-products being released. This results in a thermoset resin that combines the processing advantages offered by epoxide resins with a superior high temperature mechanical strength and good thermal stability.

The commercial version of this type of resin is a one-pack system sold as a solution in ethyl methyl ketone and is intended primarily for the preparation of reinforced laminates. Its principal features are listed below:

(a) easy to process and readily wets a wide range of reinforcements;

(b) excellent prepreg shelf stability—greater than 6 months;

(c) rapid development of high temperature mechanical strength;

(d) good thermal stability (suitable for applications requiring prolonged life at temperatures up to 200°C).

(e) low, stable values for permittivity and loss tangent;

(f) gives good high temperature bond strength to metal foils.

4.3.1 Preparation and properties of composites

The epoxy-cured Friedel–Crafts resin can be used with all forms of glass, mica and carbon fibre reinforcements in the manufacture of composites. The resin-fibre bonding of glass cloth-reinforced laminates can be improved by the use of epoxysilane or aminosilane coupling agents, resulting in an optimisation of the mechanical, thermal and electrical properties.

The precuring and pressing conditions required vary with the application but, as an indication, when prepreg (made by heating for 10–15 min at 150–160°C) is pressed at 100–500 lbf in^{-2} (7–35 MN m^{-2}) and 160–190°C, the resultant resin flow is of the order of 4%.

Postcuring is essential if the full properties of the composites are to be obtained but, since no volatiles are formed during the final cure, the conditions used are not critical. The development of high temperature mechanical strength in a glass cloth laminate pressed for 1 h at 175°C and then heated at 200°C is shown in Table 4.1.

The high temperature strength develops rapidly and after only 5 h at 200°C it reaches 75% of the room temperature strength. This value is far superior to those given by similar epoxide, epoxy novolac or silicone boards.

The thermal stability of the resin is well illustrated in Fig. 4.9, which shows the change in flexural strengths of samples of glass cloth-reinforced laminates heat aged in air at three different temperatures for up to 1000 h and then tested at 200°C.

TABLE 4.1

Postcure time at 200°C (h)	Flexural strength at 200°C	
	$lbf\,in^{-2}$	$MN\,m^{-2}$
None	25 400	175
1	48 200	332
5	60 000	414
16	62 100	428

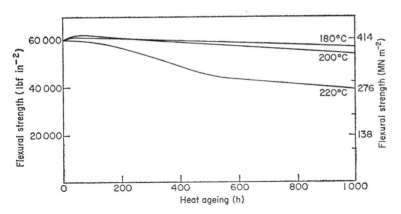

FIG. 4.9. Flexural strengths of glass cloth laminates heat aged at various temperatures and measured at 200°C.

At 180°C, 93% of the initial high temperature strength is retained, at 200°C the retention is 90% and at 220°C it is 65%. The laminates are, therefore, superior to the best type of epoxide laminates for prolonged use at temperatures up to 200°C.

Some of the other important physical properties of these glass cloth-reinforced Friedel–Crafts laminates are listed in Table 4.2.

The room temperature tensile strength is high and nearly 70% of this value is retained at 200°C, whilst the crushing and impact strengths fall into the upper range of values given by epoxide, melamine and phenolic boards. For a resin of this type the oxygen index is reasonably high and the water absorption is low, a significant factor in electrical applications.

The more important electrical properties of glass cloth laminates based on the epoxy-cured Friedel–Crafts resin are given in Table 4.3, together with a number of other postcured boards.

TABLE 4.2

PHYSICAL PROPERTIES OF THE GLASS CLOTH-REINFORCED FRIEDEL–CRAFTS LAMINATE

Properties	Value	Test method
Specific gravity	1·80	
Tensile strength at room temperature		
lbf in^{-2}	58 600	BS 2782, 301C
MN m^{-2}	404	
Tensile strength at 200°C		
lbf in^{-2}	40 800	BS 2782, 301C
MN m^{-2}	282	
Crushing strength		
lbf in^{-2}	69 500	BS 2782, 303B
MN m^{-2}	480	
Izod impact strength		
ft. lb 0·5 in	10·7	BS 2782, 306A
Flammability oxygen index	32	ASTM D2863–70
Water absorption %	0·18	ASTM D570

The best combinations of properties are shown by the silicone and Friedel–Crafts resins.

Of even greater importance is the effect of higher temperatures on the dielectric properties. A well-known feature of epoxide resin laminates is the sharp increase in both the permittivity and loss tangent that occurs above 100°C and this has limited the general use of these composites to applications below 150°C. For Class H insulation (continuous use at 180°C), silicone laminates have usually had to be used in spite of their low mechanical strengths. Figure 4.10

shows the change in permittivity and loss tangent of epoxide, silicone and Friedel–Crafts boards at temperatures up to 200°C. The Friedel–Crafts laminate shows the same stability as the silicone, in direct contrast to the behaviour of the epoxide resin laminate.

TABLE 4.3

ELECTRICAL PROPERTIES OF GLASS CLOTH LAMINATES

Property	Epoxides	Epoxy novolac	Silicone	Friedel–Crafts
Electric strength at 90°C V/mil. (V/0·001 in)	285	270	250	280
Insulation resistance MΩ	$1·7 \times 10^5$	$1·3 \times 10^5$	$2·0 \times 10^5$	$5·0 \times 10^5$
Permittivity at 1 MHz				
(a) Dry	5·05	5·34	3·80	5·00
(b) Wet	5·08	5·37	3·88	5·10
Loss tangent at 1 MHz				
(a) Dry	0·0174	0·0165	0·0030	0·0088
(b) Wet	0·0208	0·0182	0·0080	0·0091

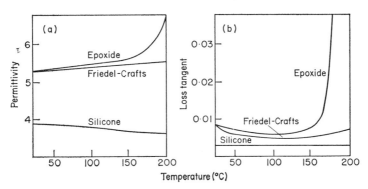

FIG. 4.10. Changes in (a) permittivity and (b) loss tangent with temperature (measured at 1000 Hz).

4.3.2 Applications

In terms of applications this resin is expected to be complementary to the hexamine-cured Friedel–Crafts polymers but, again, the electrical industry is likely to be the major user. Initial indications are that glass and various types of micaceous composites will meet the requirements of Class H insulation. Already, great interest has been shown in glass-backed mica insulation tape prepared with this resin.

Preliminary evaluations suggest that the epoxy-cured Friedel–Crafts resin will prove useful in the preparation of high performance printed circuit boards. Copper-clad glasscloth laminates are readily prepared because no volatile by-products are formed during the final resin cure. A high peel strength (10 lbf in^{-2}; 0·07 MN m^{-2}) is obtained at room temperature and 90% of this value is maintained at 200°C. In addition, the laminates remain blister-free after 6 h in boiling water, followed by 120 sec in solder at 300°C. The high-temperature mechanical strength of these composites should also prevent problems now experienced with epoxide boards, caused by the resin softening during the high speed drilling operations.

The easy processing characteristics of this resin means that thick section composites can be prepared without any difficulty and these should prove useful as structural materials (particularly in the aircraft industry), components for chemical plant and as heavy electrical insulation.

Other promising application areas include surface coatings and the preparation of carbon fibre composites, although little work has yet been carried out on these topics.

4.4 CONCLUSIONS

The Friedel–Crafts resins have now been developed to a stage where a number of products are commercially available that can be processed into composites in a similar manner to other conventional resins. These materials have a combination of thermal stability, mechanical strength and electrical properties that has led to their application in a number of industries, particularly as electrical insulation. As the demands of modern technology become more exacting these uses are likely to become even more widespread.

4.5 ACKNOWLEDGEMENT

The development work reported in this paper is the result of a team effort at the Industrial Chemicals Division of Albright and Wilson Ltd. However, the contributions made by Dr. G. I. Harris (Xylok

Resins Manager) and Messrs. F. Coxon, J. E. Golledge, B. G. Huckstepp and P. V. James deserve special mention. The author is grateful to the Albright and Wilson Division directors for permission to publish this work.

REFERENCES

1. Friedel, C. and Crafts, J. M. (1885). *Bull. Soc. Chim.*, **43**, 53.
2. Haas, H. C., Livingston, D. I. and Saunders, M. (1955). *J. Poly. Sci.*, **15**, 503.
3. Parker, D. B. V. (1956). R.A.E. Tech. Note No. Chem. 1284.
4. U.S.P. 2,870,098.
5. U.S.P. 3,405,091.
6. Doebens, J. D. and Coates, H. P. (1961). *Ind. Eng. Chem.*, **53**, 59.
7. Phillips, L. N. (1964). *Trans. J. Plastics Inst.*, **32**, 298.
8. B.P. 1,094,181.
9. B.P. 1,099, 123.
10. *Rubber and Plastics Age* (1963), **44**, 1354 (Nov.).
11. *Design and Components in Engineering* (1969), p. 15 (Nov. 5).
12. Parker, B. M. (1970). R.A.E. Tech. Rep. 70200.
13. Parker, B. M. (1972). R.A.E. Tech. Rep. 72029.
14. B.P. 1,150,203.
15. B.P. 1,305,551.
16. Harris, G. I. (1970). *British Polymer J.*, **2**, 270.
17. Harris, G. I. (1972). *The Engineer*, April 20.
18. Kondo, K. (1973). *Japan Plastics*, April 6.

CHAPTER 5

Furane Resins

A. T. RADCLIFFE
(*Quaker Oats Ltd*)

5.1 INTRODUCTION

Furane resins are not new and have been used to prepare corrosion resistant mortars for over 20 years. During this time their resistance to most acids, alkalis and organic chemicals has been well established. Early attempts to use furane resins to prepare glass fibre reinforced laminates were not successful for a number of reasons. The catalysts used were strong acids which gave a highly exothermic cure; the water of condensation evolved during cure gave unacceptable products, and also the fully cured furane resin was very brittle. The Quaker Oats Company in the USA undertook a research programme to develop resins and catalysts specifically for use in glass-reinforced plastics and the results of this work have been described in previous papers.[1] It is the intention of this paper to summarise the properties of the systems developed and to describe in more detail the progress made during the past 18 months.

5.2 CHEMISTRY OF FURANE RESIN FORMATION

Furane resins can be produced from a variety of natural products, such as corn cobs (mainly) and bagasse *via* furfural and furfuryl alcohol (Fig. 5.1).

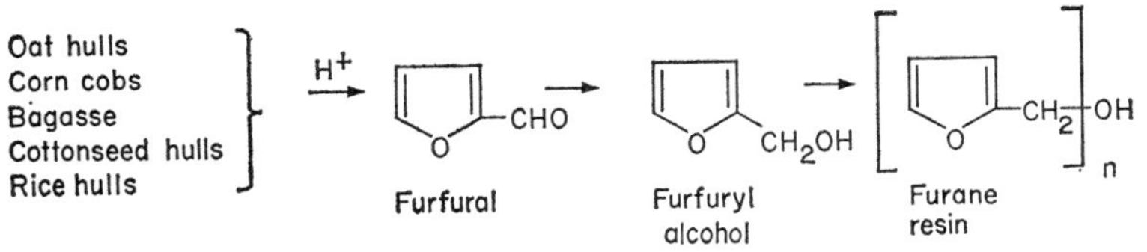

FURANE RESINS

FIG. 5.1. Sources of furane resins.

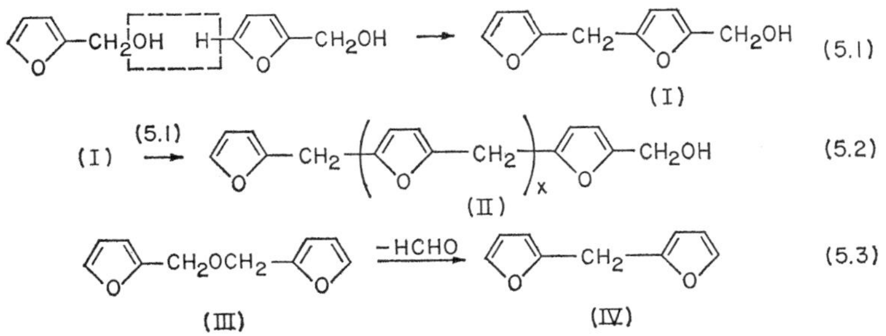

FIG. 5.2. Furane resin formation.

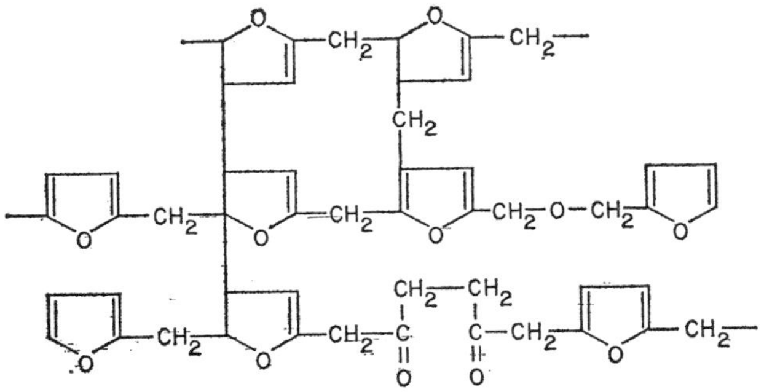

FIG. 5.3. Proposed cross-linked structure of furane resins.

The furfuryl alcohol is resinified in the presence of an acid catalyst, usually with heat, until the required degree of resinification is achieved. The polymer is normally neutralised and water is removed to give a product stable on storage.

The first step in the resinification is one of intermolecular dehydration (Fig. 5.2). This gives 5-furfurylfurfuryl alcohol (I), and continuing the reaction in the same manner (eqn 5.2) builds a continuing linear structure (II). Side reactions also occur, such as condensation to difurfuryl ether (III) and the elimination of formaldehyde forming 2-difurylmethane (IV) (eqn 5.3), among others.

These stable products are then cured with acid catalysts and heated to form highly cross-linked, infusible, resinous masses that are characterised by their outstanding chemical resistance and properties at elevated temperature (Fig. 5.3).

5.3 TRADITIONAL FURANE RESIN TYPES

For many years furane resins (uncured) have been available with a number of well-known characteristics, such as (*a*) soluble in organic solvents; (*b*) available in range from low viscosity liquids to solids; (*c*) dark coloured; (*d*) stable; (*e*) activated with acidic catalysts; (*f*) may also be heat cured to give insoluble solids without acid catalysts.

The acid catalysts which have been used traditionally with the furane resins are of the strong inorganic type, although *p*-toluene sulphonic acid has also been used.

The main applications of these resins (cured) are as follows: (*a*) chemical resistant mortars and composites; (*b*) foundry binders; (*c*) carbon–graphite binders.

Of these applications the furane resins for the foundry industry absorb some 90–95 % of the furfuryl alcohol which is manufactured today. These are usually used in conjunction with urea-formaldehyde or with phenolic resins. Only about 1–1·5 % resin is used to make the sand–resin mixture. The catalysts used are usually either phosphoric acid or *p*-toluene sulphonic acid.

5.4 LAMINATING RESINS

Because of the excellent chemical resistance shown by furane resins the Quaker Oats company undertook a research programme to develop resins which can be used to manufacture chemical-resistant glass-reinforced plastics. As a result of this research a new series of resins has been produced.

FURANE RESINS

These resins are basically furfuryl alcohol polymer, are not modified by phenolic or urea-formaldehyde resins and are liquids. The catalysts are of the modified acid catalyst type and give a rather more gentle cure than the traditional catalyst. A range of these catalysts is now available giving gel times at ambient temperature of anything from two minutes to six months. Table 5.1 gives some illustration of the development of hardness (characterised by the Barcol hardness) at various temperatures and various catalyst levels.

TABLE 5.1

BARCOL HARDNESS DEVELOPMENT DURING CURE

Barcol hardness	8% RP104B catalyst 15°C (days)	5% RP104B catalyst 25°C (days)	2·5% RP104B catalyst 40°C (days)
5	20	5	4
10	24	6	6
15	34	7	7
20	44	9	9
25	54	11	10
30	60	13	12
35	—	16	15
40	—	24	20
45	—	—	26

It is possible to accelerate the cure of furane systems by the application of heat and, after gelation, the laminate should be initially cured at 50°C and subsequently postcured at 80°C.

5.4.1 Physical properties

The physical properties of laminates fabricated from Quacorr RP100A and cured with catalyst RP104B are given in Table 5.2. The laminates were tested using ASTM and DIN test methods and were fabricated from three layers of 450 g m^{-2} ($c.$ 0·1 lb ft^{-2}) chopped strand mat and two layers of surfacing veil with a total glass content of 31%. The laminates were postcured, after gelation, for 1 h at 40°C, 1 h at 60°C and 2 h at 80°C.

TABLE 5.2
PHYSICAL PROPERTIES OF RP 100A/RP 104B GLASS-REINFORCED LAMINATES

Property	Test method	Result	Test method	Result
Specific gravity	—	1·48	—	1·48
Glass content	—	31%	—	31%
Tensile strength	ASTM D638-64T	11 400 lbf in^{-2} 78·6 MN m^{-2}	DIN	800 kg cm^{-2} (11 400 lbf in^{-2})
Elongation at break	—	—	DIN	2·25%
Tensile modulus	ASTM D638-64T	810 000 lbf in^{-2} 5590 MN m^{-2}	DIN	73 000 kg cm^{-2} 1 038 000 lbf in^{-2}
Flexural strength	ASTM D790-66	21 800 lbf in^{-2} 152 MN m^{-2}	DIN	1 532 kg cm^{-2} 21 800 lbf in^{-2}
Flexural modulus	ASTM D790-66	790 000 lbf in^{-2} 5 450 MN m^{-2}	DIN	70 500 kg cm^{-2} 1 000 000 lbf in^{-2}
Impact strength	ASTM D256-56	13·5 ft/lb in^{-1}	DIN	140 cm kg cm^{-1}
Heat distortion temperature	ASTM D648-56	218°C	DIN (Martens)	122°C
Water absorption	—	1–3%	DIN	2·5%
Coefficient of linear expansion	—	$2·4 \times 10^{-5}$ in in^{-1} °C^{-1} ($2·4 \times 10^{-5}$ cm cm^{-1} °C^{-1})	—	—
Barcol hardness	—	48–51	—	48–51
Specific resistance	—	—	DIN	$12·5 \times 10^{9}$ Ω cm
Dielectric strength	—	—	DIN	4·7 kV mm^{-1}

TABLE 5.3
RETENTION OF FLEXURAL STRENGTH AND MODULUS AFTER 1 YEAR'S EXPOSURE TO ASTM MEDIA AT ROOM TEMPERATURE

Media	Flexural strength		% Retention	Flexural modulus		% Retention
	$lbf\,in^{-2}$	$MN\,m^{-2}$		$lbf\,in^{-2}$	$MN\,m^{-2}$	
25% Sulphuric acid	15 100	104	79·5	652 000	4 500	96·2
15% Hydrochloric acid	15 100	104	79·5	695,000	4 790	102·5%
5% Nitric acid	15 300	105	80·5	606 000	4 170	89·4
25% Acetic acid	18 700	130	98·4	639 000	4 400	94·3
15% Phosphoric acid	16 000	110	84·7	674 000	4 650	99·4
5% Sodium hydroxide	17 000	117	89·5	709 000	4 890	104·6
10% Sodium carbonate	19 200	132	101·0	758 000	5 230	111·8
Saturated sodium chloride	24 800	171	130·5	883 000	6 090	130·2
5% Aluminium potassium sulphate	16 000	110	84·2	662 000	4 560	97·6
95% Ethanol	26 700	184	140·5	859 000	5 920	126·7
Ethyl acetate	21 600	148	113·7	705 000	4 860	104·0
Methyl ethyl ketone	20 500	140	107·9	666 000	4 580	98·2
Monochlorobenzene	18 900	130	99·5	689 000	4 750	101·6
Perchloroethylene	20 300	139	106·8	669 000	4 610	99·6
n-Heptane	21 800	148	114·7	720 000	4 960	106·1
Kerosene	18 600	129	97·7	561 000	3 870	82·7
Toluene	18 800	130	99·1	552 000	3 810	81·3
Demineralised water	15 100	104	79·5	621 000	4 280	91·6

The laminates were constructed from three layers of chopped strand mat and two layers of tissue mat with a glass content of 25%. The cure schedule was 1 h at 40°C, 1 h at 60°C and 2 h at 80°C.

5.4.2 Chemical resistance

The chemical resistance of furane resins, because of their long use in various composite applications, is almost legendary. In general, they were found to have good resistance to all chemicals, with the exception of concentrated sulphuric and hydrofluoric acids, strong caustic soda and oxidising media. Since the newly developed furane resins used in the manufacture of glass fibre reinforced plastics equipment

TABLE 5.4

RETENTION OF FLEXURAL STRENGTH AFTER 6 MONTHS' EXPOSURE TO ASTM MEDIA AT 65°C

Media	Flexural strength		% Retention
	$lbf\ in^{-2}$	$MN\ m^{-2}$	
Distilled water	15 700	108	80
15% Hydrochloric acid	20 200	139	104
5% Nitric acid	16 500	114	85
5% Sodium hydroxide	19 100	132	98
Sat. sodium chloride	18 800	130	96
95% Ethanol	19 400	133	99
Monochlorobenzene	21 800	150	112
Perchloroethylene	22 000	152	113
Toluene	19 700	135	101

The laminates were constructed of three layers of chopped strand mat and two layers of tissue mat with a glass content of 25%. The cure schedule was 1 h at 40°C, 1 h at 60°C and 2 h at 80°C.

have a similar chemical composition to the resin used in these composites, the same chemical resistance may be expected. Tests have shown that the new catalysts appear to optimise the chemical resistance of furane resin laminates.

Corrosion studies on these new systems have been conducted in various corrosive media at room temperature. Flexural strength retention after 1 year's exposure at room temperature and 6 months' exposure at 65°C to the ASTM list of chemicals is shown in Tables 5.3 and 5.4. Additional corrosion studies are currently being conducted with a range of even more aggressive chemicals. It is hoped to extend our information to cover the maximum service temperature of furane resin laminates in this wide range of chemicals. Preliminary results after 90 days' exposure to the chemicals are given in Table 5.5.

FURANE RESINS

The outstanding feature shown in Tables 5.3, 5.4 and 5.5 is the excellent resistance to organic media. Flexural strength retention was almost 100% in most cases and this resistance to organic media has put furane resin glass-reinforced plastics equipment into service where conventional glass fibre-reinforced plastics could not otherwise have been used.

The chemical resistance of development resin QX 300 is similar and the results after 6 months' exposure to nine of the chemicals on the ASTM list are shown in Table 5.6. The laminates were exposed at 65°C.

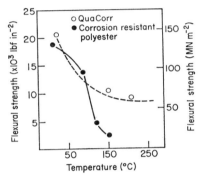

FIG. 5.4. Flexural strength of glass fibre laminates at elevated temperatures.

5.4.3 High temperature properties

The stability of furane resins at elevated temperatures is well documented in the literature. Several investigators[2-4] have conducted thermal gravimetric analyses (TGA) and/or differential thermal analysis (DTA) studies on a variety of furane resins. In each case these resins show strong thermal stability characteristics and these results, coupled with unpublished data from our own laboratories, indicate that furane resins may be used continuously at temperatures up to 150°C. Work is continuing to determine maximum operating temperature.

Gas analysis of the pyrolysis products has also been studied. Although the proportions vary with the resin type, the same relatively simple materials make up the gas stream from all furane resins (i.e. water, carbon dioxide, carbon monoxide, methane, ethane and hydrogen).

TABLE 5.5

RETENTION OF FLEXURAL STRENGTH AFTER 90 DAYS' EXPOSURE TO SELECTED MEDIA AT 65°C

Media	% Retention
30% Hydrochloric acid	75·0
60% Sulphuric acid	57·9
½% Sodium hypochlorite	61·3
10% Sodium hydroxide	63·7
20% Sodium hydroxide	52·6
10% Pyridine in water	85·0
90% Phosphoric acid	51·7
Carbon tetrachloride	106·6
Glacial acetic acid	71·7
Nitrobenzene	96·0
Ethylene dichloride	81·0
Ethyl acetate	114·4
Acetone	106·2
Benzene	124·9
Cresol	81·1
Pyridine	32·7

The laminates were constructed of three layers of chopped strand mat and two layers of tissue mat with a glass content of 25%. The cure schedule was 1 h at 40°C, 1 h at 60°C and 2 h at 80°C.

TABLE 5.6

CORROSION RESISTANCE OF DEVELOPMENT RESIN QX 300 AT 65°C AFTER 6 MONTHS' EXPOSURE

Media	% Retention of flexural strength
15% Hydrochloric acid	103·6
5% Nitric acid	84·9[a]
5% Sodium hydroxide	97·4
Sat. sodium chloride	96·4
95% Ethanol	99·4
Monochlorobenzene	111·8
Perchloroethylene	112·8
Toluene	101·0
Distilled water	80·5

[a] Indicates that QX 300 should not be recommended in this media at 65°C.

TABLE 5.7
PHYSICAL STRENGTH RETENTION OF QUACORR RP 100A AND QX 300 AT ELEVATED TEMPERATURES

Resin	Temperature (°C)	Flexural strength $lbf\,in^{-2}$	Flexural strength $MN\,m^{-2}$	Flexural modulus $lbf\,in^{-1}$	Flexural modulus $MN\,m^{-2}$	Tensile strength $lbf\,in^{-2}$	Tensile strength $MN\,m^{-2}$	Tensile modulus $lbf\,in^{-2}$	Tensile modulus $MN\,m^{-2}$
RP 100A	20	22 284	154	556 000	3 830	12 100	83	676 000	4 650
	150	10 900	75	240 000	1 650	7 900	55	482 000	3 320
	230	8 500	59	212 000	1 470	7 500	52	503 000	3 470
	260	8 288	57	248 000	1 700	7 700	53	490 000	3 380
QX 300	20	20 000	138	762 000	5 250	12 900	89	709 000	4 890
	150	11 500	79	395 000	2 720	10 800	75	581 000	4 000
	230	10 900	75	361 000	2 490	7 210	50	534 000	3 680
	260	9 200	63	369 000	2 550	9 050	63	400 000	2 760

The properties were measured by ASTM procedures using an Instron equipped with an environmental chamber. Strength measurements were made at the temperature indicated after a 15 min equilibrium time.
The laminates were constructed from three layers of chopped strand mat and two layers of tissue mat with a glass content of 25%.
The cure schedule was 1 h at 40°C, 1 h at 60°C and 2 h at 80°C.
Property improvements of 10–20% can be achieved by a further postcure at 120°C.

A further indication of the high temperature strength retention of furane resins is the heat distortion temperature which is in excess of 218°C. The retention of physical properties by QuaCorr RP 100A and QX 300 glass fibre-reinforced furane laminates at elevated temperatures is illustrated in Table 5.7 and Fig. 5.4.

5.4.4 Fire resistance

The fire resistance of laminates made from furane resins is considerably better than those made from resins such as polyesters. The inherent flame resistance of the furane resin may be demonstrated by a simple test which uses short lengths of two laminates. One

Fig. 5.5. Comparative flame resistance of (left) fire-retardant polyester resin (time, 25 sec) and (right) furane (time, 100 sec) laminates.

laminate is based on a furane resin and the other on a fire-retardant polyester resin. Each was heated by a Bunsen burner and Fig. 5.5 shows the superior behaviour of the furane laminate. The reason for this is that at elevated temperatures furane resins form large amounts of carbon; this is why such resins are used in such diverse applications as the foundry industry and as refractory binders. It is interesting to note that in such tests there appears to be little smoke evolved from the furane resins.

A number of other fire tests have been carried out on the standard resin (RP 100A) and the development fire-retardant resin QX 300. Both were tested according to BS 476 Part 6 (Fire Propagation Test) and Part 7 (Spread of Flame Test). Figure 5.6 illustrates a panel tested for spread of flame. In this test the panel edge is held against a gas radiant panel at 550°C. A fire-retardant polyester tends to burn away (Fig. 5.6), whereas a furane resin remains in place although of limited strength.

The standard resin easily achieves a Class II rating and QX 300 achieves a Class I rating. In the Fire Propagation Test (BS 476 Part 6) the standard resin RP 100A has an Index of Performance of 18–20, while QX 300 has an Index of Performance of less than 12.

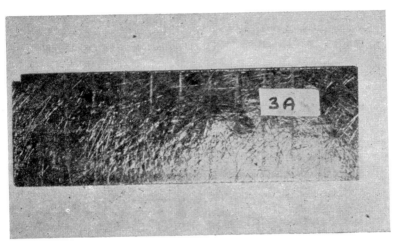

Fig. 5.6. Spread of flame test applied to a polyester resin laminate.

To obtain a Class 0 rating the Index of Performance must be less than 12 and therefore QX 300 can be classed as such a material.

Recently, a piece of fume ducting has been tested to BS 476 Part 8. This is a stringent test and is based on the testing of an actual structure under extreme conditions. During the test, hot gas was passed through the fume duct and the temperature reached was 850°C during a period of 30 min. This test was passed successfully.

The most demanding test to which glass fibre-reinforced laminates have been put is the Steiner Tunnel Test, designated ASTM E84. In this test the flame spread under forced draught conditions is related to red oak and asbestos cement board which have been arbitrarily assigned ratings of 100 and 0, respectively. Low values are most desirable and values of 25 or less are essential for many applications.

Laminates fabricated with experimental resin QX 300 (25% glass content) were recently evaluated and the flame spread rating for this system without benefit of any solid synergists was determined to be less than 25 with a fuel contribution value of 10.

5.4.5 Smoke emission

The increased concern over smoke emission has brought about comparable interest in smoke emission measurements. We have chosen the Michigan Chemical Oxygen Index and Smoke Densitometer for our laboratory evaluation of smoke emission properties of glass fibre-reinforced plastics based on furane resins. This instrument

TABLE 5.8

SMOKE DENSITY OF LAMINATES AS DETERMINED WITH MICHIGAN CHEMICAL SMOKE DENSITOMETER

Resin system	% Light transmission
Quacorr RP 100A	97
Development resin QX 300	95
Fire-retardant polyester resin	4

has been described by DiPietro et al.[5] The light source is adjusted to standardise the light transmission meter at 100% and the relative absorption by the smoke from a sample being combusted in the cylinder can be read directly from the meter.

Evaluation of the resins was conducted using oxygen levels slightly higher than the Oxygen Index previously determined for each system. This was necessary to maintain uniform combustion of the specimen which was considerably larger (2 in × 5 in × 0·125 in) (5 cm × 12·5 cm × 0·3 cm) than the normal sample for Oxygen Index determination. Laminates (25% glass) based on two furane resins and a fire-retardant polyester resin were tested by this method and the results are shown in Table 5.8.

5.5 FABRICATION OF LAMINATING RESINS

Although the basic RP 100A/RP 104B system was designed for hand lay-up, it has also been used for spray-up using the one-pot method.

Conventional equipment can be used for hand lay-up, and both Venus and Glascraft guns have been used successfully for spray-up of experimental systems now under development.

Most glass fibre reinforcements may be used with furane resins, but those having silane coupling agents are preferred. It must be remembered that the glass fibre binder system was developed for use with polyester resins. It is more soluble in the styrene used in these resins than in the furane resin system. The furane resins wet the glass more slowly and there is a great temptation to add too much resin. If this is done and the glass content is of the order of 15%, resin-rich layers may result which are liable to crack on flexing. The glass contents of a satisfactory laminate should be between 25 and 35%.

The laminates can be released from the mould by using polyethylene terephthalate film (Mylar, Melinex), PTFE (Fluon, Teflon), or Carnauba wax. The wax release agent tends to stay with the laminate when it is stripped and, therefore, metal moulds should be re-waxed soon after use to prevent corrosion. Many of the release agents used for polyester resins are not suitable for use with furane resins. Examples are polyvinyl alcohol, silicones, particulate graphite and stearates.

The gel time of the catalysed resin can be made to vary between 4 min and 5 days and, by proper choice of catalyst type and amount, almost any desired cure cycle can be obtained. Laminates may normally be stripped from the mould after 16 h at 22°C. The stripping time is considerably influenced by the nature of the mould and the ambient temperature conditions, whereas the pot-life of the resin is controlled more by the resin temperature. With a mould which acts as a heat sink and when the ambient temperature is lower than 20°C, the cure should be accelerated by the input of heat rather than by increasing the catalyst percentage. When using elevated temperature curing it is important that the laminate is allowed to gel at ambient temperature before applying heat and that an initial temperature of 50°C is not exceeded.

As furane resins do not require a resin-rich surface for good chemical resistance, it is not necessary to have a gel coat. In fact, unreinforced resin layers are brittle and may crack in flexure if more than 0·25 mm (0·01 in) thick. The cure of furane laminates is not air inhibited as are polyester resins and, therefore, it is not necessary to incorporate wax in the final coat.

Furane resins may be used alone to fabricate chemical resistant equipment but may also be used as the chemical resistant barrier in composite constructions with other resins. This system shows advantage when large structures having a high raw material cost are being made and when a cheap general-purpose polyester can be used as the second resin. Furane chemical resistant barriers applied by hand or spray-up have been used successfully inside filament wound polyester tanks and pipes. This composite system is less attractive with smaller items, or when overspill of the contents of the tank could attack the polyester backing layer.

5.6 APPLICATIONS

Furane resin laminates will find application in several areas where corrosion resistance is required. For example: (a) in situations where the corrosion is associated with organic media such as solvents; (b) in situations where corrosion is encountered above temperatures of 110°C, i.e. fume stacks; (c) in situations where fire retardance as well as corrosion resistance is required, i.e. fume ducts carrying organic solvents.

The low smoke emission would be an added benefit for the latter application.

5.6.1 Case histories

Since the new furane laminating resins were introduced, a considerable number of interesting case histories have been established. They demonstrate the fact that these resins are no longer just laboratory curiosities. These case histories include the following:

(1) A tank which forms part of an effluent processing plant in a UK pharmaceutical factory and which measures 3 m diameter, 3 m high. It is required to operate at temperatures of from 20 to 40°C and under pH conditions of from 2 to 14. Sodium hydroxide, hydrochloric acid, petroleum ether, pyridine, benzaldehyde, ethanol, chlorinated solvents, aromatic solvents and water may all be encountered.

(2) An even larger tank (8 m diameter, 5 m high) has been operating successfully, since November 1972, handling hydrochloric acid and chlorobenzene at a temperature of 65°C. Its use illustrates one of the major advantages of furane resins, i.e.

their ability to handle a combination of corrosive and organic solvents.

(3) A combination of techniques has been used in the USA to manufacture a fume stack which measures 1·5 m diameter, 16 m long. The inner layers were built up by the hand lay-up technique and the outer layers applied by filament winding. The stack is required to handle sulphuric acid, hydrochloric acid, sulphur dioxide and trioxide and fly-ash. In this application furane resins were chosen because the temperatures can occasionally reach 193°C. A normal operating temperature is, however, 82°C. A previous stack made of stainless steel failed in six months; the furane-based stack has been in use for at least one year to this date.

(4) A tank, 2·20 m diameter, 2·25 m high, fabricated from a furane laminate and operating at 30°C in a mixture containing dilute hydrochloric acid, trichloroethylene and trichlorethane. This was installed in January 1973.

(5) Pump impellers, 0·5 m diameter, fabricated from a furane/polyester composite and operating at 92°C in a mixture containing hydrochloric acid, chlorobenzene, ethyl benzene and benzene. The first impeller was put into service in November 1971.

(6) Two chemical resistant polyester tanks, 3 m diameter, 5 m high, containing the above mixture showed signs of failure and were relined with furane laminate in January 1973.

(7) A tank, 1·82 m diameter, 1·82 m high, fabricated from a furane/polyester composite and operating at 60°C in a mixture of dodecyl benzene, Shellsol A and sulphuric acid. In service November 1972.

(8) A scrubber, 14 m high, 3·8 m diameter, fabricated from a furane/polyester composite and operating at 80°C in a mixture of 30–40% sulphuric acid, 1% nitric acid and 1% dinitrotoluene. In service May 1972.

(9) A 3000 gal cone-shaped separator fabricated from a furane laminate and operating at 100°C in a mixture of pine oil (terpenes) and aqueous calcium salts. In service January 1973.

(10) A fume ducting fabricated from a furane laminate and operating at 90°C in a mixture of 20–30% hydrochloric acid, chlorobenzene and chlorinated phenols.

(11) A tank, 7 m high, 5 m diameter, fabricated from a furane/polyester composite and operating at 80°C in a mixture of hydrochloric acid and chlorobenzene.

(12) Two tanks, 1·5 m diameter, 2·5 m long, fabricated from a furane/polyester composite and operating at ambient temperature in a mixture of hydrochloric acid, sulphuric acid, butyl alcohol and octyl alcohol.

(13) A tank, 1 m diameter, 2 m long, fabricated from a furane/polyester composite and operating at ambient temperature in a mixture of hydrochloric acid and xylene.

(14) A pipeline, 40 m long, 0·2 m diameter, fabricated from a furane laminate operating at 65°C in a mixture of tetrahydrofuran, formic acid, methanol, methylene chloride, chloroform and trichloroethylene.

(15) Nine tanks fabricated from a furane laminate and operating at ambient temperature in a mixture of carbon disulphide, sodium sulphate and sulphuric acid. In service June 1972.

(16) Two 11 000 gal tanks fabricated from a furane laminate and operating at 90°C in a mixture of sodium bicarbonate, sodium chloride, hydrochloric acid, methanol, ammonium chloride, sodium hydroxide, sulphuric acid, sodium sulphate, isopropanol and dimethyl dialkyl ammonium chloride. In service June 1972.

5.7 CONCLUSION

Hopefully, this paper has indicated that the new furane resin systems can be successfully used to fabricate glass-reinforced plastics equipment capable of service in a wide variety of corrosive environments. The use of furane resins should enable the glass-reinforced plastics industry to fabricate equipment to serve in conditions for which the previously available resins were inadequate.

5.8 ACKNOWLEDGEMENT

I should like to thank my colleagues at The Quaker Oats Company in Chicago and Barrington, USA, whose help has made this paper possible.

REFERENCES

1. (*a*) Furane Resin Laminates—A Unique Combination of Properties: Bozer, K., Brown, L. and Watson, D. (1971). 26*th Annual Conference, S.P.I.;* Lens, T. and Radcliffe, A. (1972). *B.P.F. Reinforced Plastics Conference.* (*b*) High Temperature and Combustion Properties of Furane Composites: Bozer, K. and Brown, L. (1972). 27*th Annual Conference, S.P.I.*
2. Tech. Doc. Rep. No. WADD TR 61–72, Vol. XV, 'Alumina Condensed Furfuryl Alcohol Resins'.
3. Werwerka, E. M., Walters, K. L. and Moore, R. H. (1969). *Carbon*, **7**, 129.
4. Fitzer, E., Schaefer, W. and Yamada, S. (1969). *Carbon*, **7**, 643.
5. DiPietro, J., Barda, H. and Stepniczka, H. (1971). *Textile Chemist and Colorist*, **3**, No. 2, 45.

CHAPTER 6

Recent Developments in Polyurethanes

A. BARNATT

(*Lankro Chemicals Ltd*)

6.1 INTRODUCTION

In 1937 Otto Bayer and his team of chemists discovered the reactions leading to the formation of polyurethanes, but it was not until about 1950 that his discovery was put to any significant commercial use. Since then, polyurethanes have enjoyed particularly rapid growth, such that the consumption in Western Europe has risen by 16% p.a. to a 1972 figure of 552 000 tons,[1] which was distributed as in Table 6.1.

TABLE 6.1

DISTRIBUTION OF POLYURETHANE CONSUMPTION IN WESTERN EUROPE

Application	% of total consumption
Flexible slabstock	53
Flexible moulding	9
Rigid foam (refrigerators)	12
Rigid foam laminates	2
Rigid foam slabstock	2
Rigid foam (miscellaneous)	8
Semi-rigid foam	3
Microcellular shoe soles	7
Miscellaneous elastomers	4
Total	100

Provided that present raw material problems are solved growth is expected to continue at an overall rate of about 11% p.a., causing the market to top the million ton level by 1980. Flexible foam is expected to grow at about 10% p.a., while rigid foams should reach 15%. Considerable growth is expected in the footwear industry and in the use of microcellular elastomers for motor car bumpers.

Technical development in polyurethanes has progressed faster in some areas than in others, and the distribution of polyurethanes in the applications shown in Table 6.1 is changing. This paper, after briefly describing the way in which polyurethanes are formed, outlines the type of developments which are causing this change.

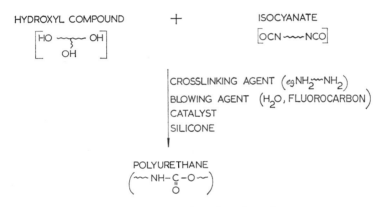

FIG. 6.1. The formation of a polyurethane.

6.2 FORMATION OF POLYURETHANES

The general principles of urethane formation are well known and are based on the reaction of an isocyanate with a compound containing an active hydrogen. Many formulations are in commercial operation and it is not the intention to discuss them in this paper. However, it is of interest to outline some general principles in order to show the wide range of products available from polyurethane.

Figure 6.1 illustrates, in a very basic manner, the type of formulation used to manufacture polyurethanes. Not all the constituents are necessary in order to make a simple polyurethane, but, conversely, many formulations employ mixtures of polyols, cross-linking agents, catalysts, blowing agents, etc. For example, a conventional flexible foam can be produced by using a 3500 MW triol as the hydroxyl compound, water (which reacts with isocyanate to form carbon dioxide) as the blowing agent, toluene diisocyanate (TDI) as the isocyanate, together with appropriate catalyst and silicone. For a rigid foam it would be desirable to use a polyol, and usually isocyanate, of greater functionality, so that a more cross-linked and rigid polymer structure would be formed. Intermediate-type foams can be made by variation of the hydroxyl and isocyanate components, a good example being the semi-rigid polyurethane foam used in the majority of motor car crashpads. Yet another variation is to omit

TABLE 6.2

CLASSIFICATION OF POLYURETHANE MATERIALS

Cellular	*Solid*
Flexible	Elastomers
Rigid ⎫ Homogeneous structure and	Coatings
Semi-rigid ⎭ integral skin	Rigid
Elastomeric (microcellular)	Adhesives

the blowing agent, when a solid polyurethane is produced, which, according to the choice of other constituents of the formulation, may be rigid or elastomeric. An intermediate cellular stage, which is resulting in a significant growth area for polyurethanes, is to significantly reduce the quantity of blowing agent, such that a microcellular structure, rather than a foam, is produced.

Bearing in mind the vast amount of formulation variation possible, it is convenient to classify the types of polyurethane which are currently available in commercial quantities (Table 6.2).

Of these types of polyurethane, flexible foam manufacture is a well-established process which accounts for a very large percentage of total polyurethane tonnage. Change has been in the form of steady development rather than sudden breakthrough. Rigid and semi-rigid foam production is likewise well established, although significant change is beginning to take place on the integral skin versions. Microcellular polyurethane elastomers are a comparatively new field. Significant progress is being made in the footwear and automobile industries, both of which are considered to be major growth areas for the next two or three years. The solid, or unblown, polyurethanes currently account for only a small percentage of polyurethane consumption. Possibly they are at an earlier stage of development than the cellular materials, and one would hope that their use will significantly increase in the future.

As already suggested, it is a change in the relative importance of these types of polyurethane which is likely to be a result of development.

6.3 APPLICATIONS

6.3.1 The motor car and transport industries

Traditionally motor cars have used semi-rigid foams for filling crashpads and flexible foam for seating, but new applications have been found due to the development of integral skin semi-rigid foam, cold cure flexible foam and microcellular bumper compounds.

Integral skin semi-rigid foam

Although this type of foam has been known for some years it is only recently that it has gained commercial significance, particularly in the UK. This has largely been because foams did not have adequate physical properties without the use of cross-linking agents with possible carcinogenic properties. Systems developed over the last two or three years use alternative polyols and cross-linking agents, and possess physical properties adequate for most applications.

The integral skin foam is formed by using a low boiling point liquid (usually a fluorocarbon such as Refrigerant 11) as the blowing agent. During the polyurethane manufacture, heat is evolved, which vaporises the low boiling point liquid and produces a cellular, or foamed, structure. If, however, the foam is produced in a mould, particularly a cold one, the amount of vaporisation near to the surface of the mould is less than that at the centre. This produces less expansion at the surface of the moulding, hence forming a urethane skin of higher density than the rest of the moulding (Fig. 6.2).

Examples of the use of integral skin semi-rigid foam in cars are numerous, and Fig. 6.3 illustrates one of the possible applications.

Fig. 6.2. Comparison of water blown and Refrigerant 11 blown semi-rigid polyurethane foam. The foam with the skin has been blown using Refrigerant 11.

Fig. 6.3. Pilot scale seat manufactured from integral skin semi-rigid polyurethane foam (courtesy of Joseph Lucas, Birmingham).

Cold cure foam

Another fairly recent polyurethane development influencing the motor car industry is the so-called cold cure foam. Like integral skin semi-rigid foam, cold cure foams have been known for a number of years, but only recently have they started to make any commercial impact (particularly in the UK). Their introduction gives immediate advantages to both the method of manufacture and the properties of the moulded cushion.

Traditionally, seat mouldings were manufactured by a hot cure process which demanded that the foam, whilst still in a metal mould, was subjected to a heat cure cycle. This required large ovens and long conveyors and, hence, elaborate and expensive equipment. On the other hand, cold cure foams, as their name suggests, do not require a heat cure. They can be produced in much cheaper moulds and, hence, require less capital outlay on equipment.

The 'cold cure' properties of the foam are achieved by using more reactive polyols than are used in hot cure formulations, giving more evolution of heat during urethane formation, which cures the foam sufficiently for it to be demoulded. The final stage of cure is achieved after the cushion is removed from the mould, and is usually complete after a few days' storage at room temperature.

The advantages of cold cure foam over hot cure foam can be summarised as: (a) lower hysteresis loss; (b) improved sag factor; (c) improved fatigue resistance; (d) improved flammability properties.

Hysteresis is a measurement of the energy lost or absorbed by a foam when subjected to deformation. Foams with a high hysteresis show poor resilience and can be slow in regaining their original shape after deformation. Figure 6.4 shows typical hysteresis curves for hot and cold cure foams. The hysteresis curve for cold cure foam resembles that for rubber latex and is the main reason why cold cure foam is frequently referred to as 'latex-like foam'.

The *sag factor* of a foam is the ratio of the loadings required to compress the foam to 65% and 25% deflection. A foam with a high sag factor is usually more comfortable than one with a low sag factor, since the load required for initial deformation is low, while the higher load for 65% deformation provides the necessary firmness for support. Cold cure foams usually have a sag factor of about 3, compared to a value of about 2 for hot cure foams.

The *fatigue resistance* of a flexible foam is a measure of its durability in service. Various methods have been used to comr hot and cold cure foams, all of which agree that cold cur better material.

The *flammability characteristics* of these highly resi¹' foams show a distinct improvement over their hr parts. Most highly resilient foams satisfy M Standard 302 specification, a test rapidly g American and European motor car ind· been put forward regarding this differ

istics, the most widely accepted being that the melt phenomena are different. High resilience foams tend to melt and drip away from the flame more quickly than hot cure foams, with obvious advantages in preventing initiation of a fire.

An added advantage of cold cure foams is that, because of the lack of need for a heat cure cycle, they can be moulded directly into vinyl or ABS skins without skin deformation, hence producing a covered seat direct from the mould.

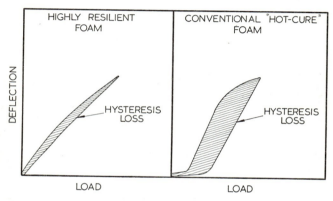

FIG. 6.4. Typical hysteresis curves for hot cure and cold cure foams.

Bumper systems

... the most significant development of polyurethane in the ... industry is in bumper bar applications. During an evalua... in the USA of 33 different plastics and plastics com... two were not damaged by a sharp-impact test. ...ocellular polyurethane and a combination of an ...d with a semi-rigid polyurethane foam.[2] ... type of test is by no means the only reason for ...owth into bumper applications, the overall ...; (b) compressibility with minimum defor... load/deflection curve; (d) high energy ...mperature range.
...ssed by solid urethane elastomers, ...d solid materials have the com-
...ly be described as 'toughness'.
...h account for the success of

microcellular polyurethane, in turn it is clear that, first, the lightness arising because of the low density of the moulded bumper (frequently about 0·6 g cm^{-3}) gives obvious advantages in both cost and design.

Secondly, the property of compressibility with minimum deformation arises because of the microcellular structure of the material. Very little lateral deformation is noticed until all the air space in the structure has been crushed, and this characteristic is of value in those applications where the elastomer is used in a confined space to absorb shock.

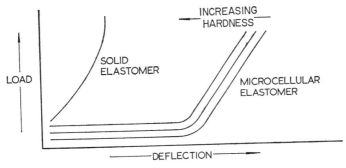

FIG. 6.5. Load/deflection curves for solid and microcellular polyurethane.

Thirdly, the shape of the load deflection curve shows a major difference between microcellular and solid elastomers (Fig. 6.5). At up to about 40% deflection the load/deflection curve is approximately linear, and then becomes progressively steeper. The practical application of this property is that impacts are absorbed by the microcellular polyurethanes, which recover their original shape after the source of impact has been removed. This is in contrast to a metal bumper, where transmission of impact force and permanent damage are common.

One of the major developments of polyurethanes in cars came in the USA in 1968, when the Pontiac GTP used this material in the front bumper. The bumper was able to withstand an impact test of 4000 lb (1814 kg) moving at 4 m.p.h., and Pontiac continue to use this material, e.g. in the 1973 Grand AM.

The 1974 Pontiac Firebird uses microcellular polyurethane as the complete energy absorbing unit at both the front and rear of the car, no other shock absorbing method being used (Fig. 6.6).[3]

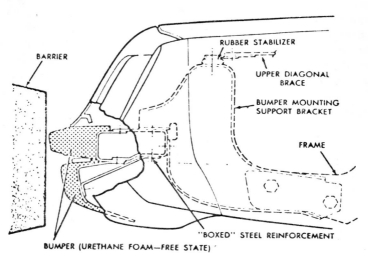

BEFORE IMPACT

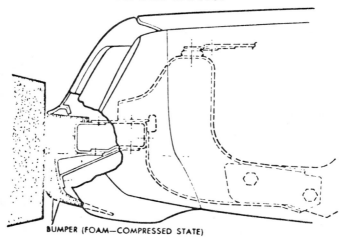

DURING IMPACT

FIG. 6.6. 1974 Pontiac Firebird front bumper system (courtesy of *Rubber Age*).

Another interesting use of microcellular polyurethane is the experimental safety car developed by a joint Pirelli/Fiat team.[4] This car has nearly 1 ft (0·31 m) thickness of elastomer at both front and rear, and is currently receiving much attention from motor vehicle safety experts in Washington.

A recent development in the UK is to equip the export model Marina with polyurethane bumper ends. Another US bumper system, outlined recently in the *Sunday Times*, is a rectangular 'sausage' of polyurethane filled with air at about 8 lbf in^{-2} (0·06 MN m^{-2}). This design is claimed to protect cars adequately at impact speeds up to 20 m.p.h. An alternative design filled the polyurethane skin with water instead of air, but this was rejected because of weight problems, the need to refill after impact damage and the possibility of the water freezing in winter.

The high energy absorbing properties over a wide temperature range are the fourth reason for growth of polyurethane into transport. Although it is well known that solid polyurethane cannot withstand excesssive vibration due to heat build up and premature breakdown, this is not the case with applications such as engine and suspension mountings.

A good example of the utilisation of absorption properties is the use of 'Microvon' microcellular polyurethane in the Vickers battle tank.[5] Bump stops of this material withstand loads up to 2300 lb (1000 kg) on a 127 mm diameter disc, 51 mm deep. Auxiliary springs on the vehicle also use microcellular polyurethane which, because of its low lateral deformation under load, can easily be accommodated in confined spaces.

Miscellaneous

Another possible use for polyurethane in cars is to fill the petrol tank with reticulated polyester foam, which is said to reduce the fire hazard in a crash by preventing explosion or rapid spreading of the petrol if the tank is punctured. Typically, the foam would fill about 3% of the fuel tank volume.

6.4 THE FURNITURE INDUSTRY

The furniture industry is already well used to polyurethanes, since it is a very large consumer of flexible foam for bedding and upholstery and low density rigid foam for chairshells. Changes have occurred in the type of flexible foam, in that a large amount is now the cold cure type already discussed in connection with the motor industry. These traditional applications have the common factor that the foam is not seen in the finished article. The rigid foam chairshell is covered in flexible foam and this, in turn, is covered with a furnishing fabric.

The change which has already started is that polyurethanes are being used for items of furniture which can be seen, a situation due to the introduction of integral skin structural rigid foam and polyurethane coated fabrics.

6.4.1 Structural rigid foam

Structural foam has been known, in a less refined form, for some years, but its recent introduction into furniture has been caused by three main factors: (*a*) wood has become in short supply, (*b*) foam systems have been developed to give physical properties adequate for the stringent requirements of articles such as chair legs; (*c*) foam systems have been developed to give much faster demould times than previously—this particularly affects the economics of moulding polyurethanes.

Traditionally, great emphasis was placed on the suitability of polyurethanes for short production runs of items which, when produced by other means, had a high labour content or were produced from expensive raw materials (e.g. wood). In such cases the high capital investment for thermoplastic mouldings was not justified. However, the introduction of polyurethane foam systems with fast demould times has changed this situation radically, and it is now frequently economically attractive to use polyurethanes for long production runs of large items and simple shapes.

The latest foam systems have very strong skins with low density centres. They can be used on visible surfaces of furniture, and produce slim mouldings with excellent rigidity and low weight. This type of polyurethane is already established in Europe and, it is expected, will be used in significant quantities in similar applications in the UK within the next two years. Figures 6.7 and 6.8 show samples of the type of mouldings which can be made using integral skin structural rigid polyurethane foam.

6.4.2 Urethane coated fabrics

Of the many types of fabric coating available today, polyurethane stands out because of its special suitability for the coating of textile fabrics, woven materials, fleeces and every type of non-woven material. Although polyurethanes for fabric coating do not all fit exactly into the category of thermoplastic materials, they are certainly one of the fast developing areas of polyurethane technology and, as such, are worthy of brief mention.

Fig. 6.7. Coffee table in rigid urethane foam (courtesy of Moulded Furniture Products (Appollo Chair Company), Ashington).

Polyurethane coated materials distinguish themselves by properties such as good abrasion resistance, excellent adhesion between coating and substrate, good low temperature flexibility and good resistance to dry cleaning. In addition to this they have good handle and can be formulated to feel very similar to leather. A big advantage over PVC coated fabrics is a degree of water vapour permeability, with the obvious advantages in footwear, garments and upholstery. These materials appeared on the market about 3 years ago, and have gained acceptance to the stage where Western Europe manufactured about 100m m^2 in 1973.

The usual techniques of applying polyurethane to a fabric are either direct coating or transfer coating from a release paper, both methods being well known for coating other materials. Application can be from either one-component or two-component systems. For the one-component system the fully reacted polyurethane is produced by the chemical manufacturer, whereas for the two-component system the polyurethane is produced *in situ* either directly on the

FIG. 6.8. A urethane door on the Tricity Tudor Refrigerator (courtesy of Thorn Domestic Appliances (Electrical) Ltd).

fabric or on the release paper. Because of processing problems, two-component systems are being replaced by one-component systems for transfer coating. For direct coating, however, the two-component systems are still used to a significant extent, the types of substrate frequently used in direct coating being unable to withstand the solvents and drying temperatures usually required for one-component systems.

6.5 THE REFRIGERATION INDUSTRY

About half the rigid foam produced in Western Europe is used for insulation of refrigerators, mainly because it has twice the insulating value of any other commercially available insulating material, combined with a high strength to weight ratio. Developments in this area are likely to be a steady improvement in the physical properties of the foam, a decrease in the foam density and a move to polyurethane from other insulating materials. It is not intended to dwell on this type of development in this paper, but rather to concentrate on those areas where more rapid change is likely.

6.6 THE FOOTWEAR INDUSTRY

The footwear industry represents a major opportunity for the growth of polyurethanes into a new type of market. There are about 200m shoes manufactured annually in the UK, most of which have stuck-on rubber or PVC soles.[6] They require over 20 000 tons of shoe sole material.

Leather was traditionally used for shoe soles, followed by rubber which was introduced about 40 years ago. A significant change came about 1950, with the introduction of resin rubber, which looked like leather but was cheaper and had improved wear properties. About five years later a major processing advance came with the technique of attaching soles to the uppers by a moulded-on technique. The use of rubber grew rapidly and reached a peak in the mid-1960s when PVC with its advantage of easy processing, was introduced to the footwear market. PVC rapidly gained acceptance as a shoe sole material and is now used on about 25% of shoes manufactured in the UK.

Polyurethanes made their appearance in the late 1960s. It did not achieve its anticipated growth rate in the UK, although the picture is more promising in Europe. It has been forecast that polyurethanes will take up the demand for extra shoe soles at the expense of PVC and that this year polyurethanes will grow at a rate of 150% p.a. in the EFTA countries, 500% in France and the Benelux countries, 300% p.a. in Germany and 200% p.a. in Italy.[7] In Europe between 7m and 8m pairs of shoes a month are fitted with polyurethane soles and this year the US market is estimated at about 5m pairs.[8]

The current advantages of polyurethanes over other soling materials are: (*a*) good properties; (*b*) versatility; (*c*) lower densities.

Good properties—These include wear resistance, resilience, thermal insulation, slip resistance and oil resistance.

Versatility—This enables polyurethanes to be used for *in situ* soling, unit soling and a wide variety of sole design. Different grades are available to meet different quality requirements.

Lower densities—Lower densities are available than were previously possible with polyurethanes.

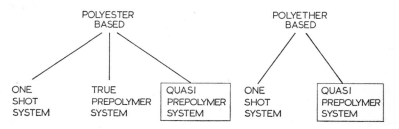

FIG. 6.9. Classification of polyurethane shoe sole systems.

This list of advantages should ensure the future of polyurethane in the shoe sole market and furthermore, answers have now been provided to earlier criticisms, processing problems and high scrap rates.

Polyurethane shoe sole materials can be either polyester or polyether based. In general, the polyester based systems give better wear properties and are used for applications such as industrial footwear and children's footwear. The polyether based systems have wear properties which do not match those of polyester based systems, and tend to be used for ladies' footwear and applications which have a shorter life due to fashion changes. It must not be assumed that polyester based systems are necessary for all applications. This is not the case, and there is a place for both types of system in the shoe sole market.

Figure 6.9 shows the types of polyurethane shoe sole system which are commonly known. As can be seen in the figure, three types of polyester system and two types of polyether system are commonly known. However, by far the greatest amount of success has been achieved with the quasi-prepolymer systems and it is these that are now making very significant inroads into the footwear industry.

6.6.1 Polyether based systems

Both integral skin and homogeneous structure polyether systems are commercially available. We at Lankro believe the homogeneous-type structure to have distinct advantages over the integral skin structure, for the following reasons:

(1) The wear properties of the sole are constant throughout its life, and wear does not accelerate rapidly when the skin has worn through.
(2) Localised skin damage in an integral skin structure, which can occur at a very early stage of the life of the shoe, can have serious effects on the wear properties of the sole.
(3) The water absorption of the low density core of an integral skin system is usually high, and localised skin damage allows water to penetrate into the bulk of the sole.
(4) Flex properties of polyurethane, particularly at low temperatures, tend to decrease as the density increases. The high density skin of an integral skin system is therefore a potential failure area for low temperature flex cracking.

6.6.2 Polyester based systems

Reasons for the somewhat slow progress of polyester based systems to date have been: (*a*) shelf life problems with the prepolymer; (*b*) the very fine limits on the mixing ratio required to produce acceptable shoe soles.

By using specially designed polyesters and by employing stabilisation techniques during manufacture of the prepolymer, both of these problems have been solved. It is believed that this has provided the shoe sole industry with a very significant advance in polyurethane technology.

Figure 6.10 shows viscosity/time graphs for stabilised and unstabilised prepolymer. The practical advantages of a stabilised prepolymer proved to be enormous. It could be left without fear of degradation in heated machine tanks for prolonged periods, and it was no longer necessary to adopt ultra-cautious premelting procedures prior to pouring the prepolymer into the shoe sole machine.

The latest polyester shoe sole systems are designed to give acceptable shoe sole mouldings over an operating mixing ratio of $\pm 5\%$ of optimum, in contrast to earlier systems which require a limit of at least $\pm 1\%$ (Fig. 6.11).

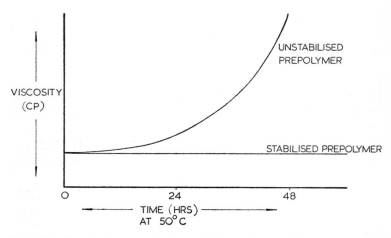

FIG. 6.10. Viscosity/time graphs for stabilised and unstabilised urethane shoe sole prepolymers.

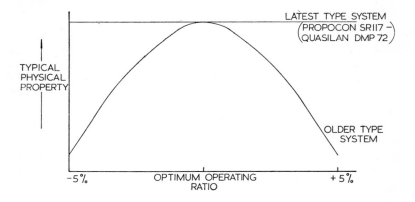

FIG. 6.11. Operating ratio/physical property graphs for old and new type polyester shoe sole systems.

6.6.3 Machinery

Machinery for shoe sole production, whether with polyether or polyester based systems, tends to be complicated and expensive. Less complicated machinery is in commercial use for polyether based systems, which are usually processed at room temperature, and the advantages and disadvantages of the various types of machinery are a matter of continual debate. Figure 6.12 shows a typical shoe sole machine, capable of producing either polyester or polyether based unit soles.

FIG. 6.12. A shoe sole machine capable of producing polyester and polyether systems (courtesy of B.U.S.M. (Leicester)).

6.7 MISCELLANEOUS APPLICATIONS

In addition to the developments already discussed there are a number of miscellaneous developments which, if successful, could lead to very significant use of polyurethanes.

6.7.1 Hydrophilic foams
By special design of the polyol component of the urethane formulation it is possible to produce a flexible foam which absorbs water. At this stage little commercial use has been made of this technology, but one cannot help but consider the large markets available should hydrophilic foam replace such materials as cottonwool and sponges.

6.7.2 Packaging
Packaging with polyurethane foam at present holds about 5% of the Western European flexible foam market.[1] This particular application

represents a potentially large volume sector, where the advantages of polyurethane are flexibility, non-abrasive properties, non-slip properties, light weight, softness, strength and good thermal insulation.

A recent development in this field is the use of polyurethane foam in postal bags, as a replacement for the conventional cotton waste and waste paper lining. This has given significant weight reductions in the package, with resultant reduction in postal charges.

6.7.3 Running tracks

An interesting development of the last two years is the use of polyurethane for running tracks. In one system the polyurethane components are mixed, poured into position and left to cure for about half an hour. A thin layer of polyurethane, containing $\frac{1}{2}$ in (12·7 mm) long nylon fibres is then applied and the fully cured-up coating system is said to be a substitute for grass.

6.7.4 Road insulation

The following is a quotation from *Scrap and Waste Reclamation and Disposal*, December 1970:

> Plastic foam is replacing the gravel beds traditionally used beneath German motorways for frost insulation and for load-bearing. Use of the rigid polyurethane foam has led to considerable saving. The traditional gravel bed was 2 ft (0·61m) thick. During the past few years the costs of excavating the gravel and transporting it has become prohibitive. Some 2000 lorry-loads of gravel are needed for each mile of motorway. Now the foam is sprayed onto the road at a rate of 100 ft^2 (9·3m^2) per minute. When this hardens into a layer $1\frac{1}{4}$ in (32 mm) thick, the top course of the road is applied directly onto the foam. About $38\frac{1}{2}$ tons of chemicals are needed for each mile and these can easily be brought to the site by five road tankers.

Polyurethane has also been put to similar use in Scandinavia but no UK uses have been noted to date. The tonnage possibilities for this type of application are enormous.

REFERENCES

1. Tunbridge, T. (1973). Polyurethanes, outstanding growth, *Europlastics Monthly*, February.
2. Cacciotti, P. J., Jr. (1973). Microcellular urethanes in automotive exterior parts. *J. Elastoplastics*, **5**, 74.

3. *Rubber Age*, October 1973, 38.
4. *Plastics and Rubber Weekly*, August 18, 1972, No. 442, 13.
5. *Plastics and Rubber Weekly*, November 17, 1972, No. 455, 23.
6. Hall, E. F. (1970). Footwear machinery: trends and developments, *Rubber Journal*, April, 55.
7. Outlook good for shoe trade applications, *Modern Plastics International*, November 1971, 18.
8. Urethane takes giant steps in fancy footwear, *Chemical Weekly*, October 18, 1972, 23.

CHAPTER 7

Processing Equipment for Cellular Polyurethanes

J. B. BLACKWELL
(*Viking Engineering Co. Ltd*)

7.1 INTRODUCTION

Probably the most significant factor that differentiates urethane equipment from other plastics manufacturing machinery is that the polymer is made for the first time in the end-user's factory and not in the factory of the raw materials supplier. In the case of other plastics machinery, the end-user would normally apply pressure, vacuum or heat, or combinations of these physical factors, to shape or process the polymer. With polyurethanes, we are initially dealing with liquids that must be: (*a*) stored in temperature-controlled conditions; (*b*) accurately metered; (*c*) mixed for a very short time, as a reaction normally commences after 10 sec; (*d*) fed into a mould or conveying system where they are subsequently cured either by the exothermic reaction or by application of external heat.

This paper will deal with the normally accepted method of storage, temperature control, metering and mixing for the two main materials, that is the polyol and the isocyanate. Subsequently, I will deal with the recent developments in the areas of flexible foam and rigid foam application technology.

7.2 RAW MATERIAL STORAGE

The raw materials may be supplied in bulk containers or in drums. If they are supplied in drums then they may be tipped directly into

PROCESSING EQUIPMENT FOR CELLULAR POLYURETHANES

the machine. If they are supplied by road tanker then the raw materials must be transferred and stored in a bulk storage system.

Figure 7.1 illustrates a typical bulk storage system for the polyol (the isocyanate system is somewhat similar). The liquids are transferred via a pump to the bulk storage tank which is fitted with silica gel breathers, a level indicator and a material temperature thermometer. Since control of temperature is so important a heat exchanger is usually fitted.

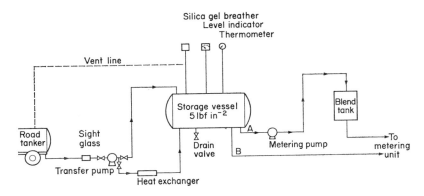

FIG. 7.1. Typical flowline diagram for a polyol bulk storage system.

If blending equipment is to be supplied for the adding of catalyst or activators, the whole process can be made automatic by the use of accurate metering pumps fitted with predetermining counters which feed the main materials to the blending tank.

7.3 METERING UNITS

The metering units can be divided into two main systems, namely low pressure systems and high pressure systems. However, the one essential feature of any metering unit is an accurate metering pump. For every revolution that a pump makes it is essential that the output is constant. Flow meters can be used to measure pump output but, in my opinion, accuracy is of extreme importance, as will be demonstrated subsequently.

A further problem which is encountered with isocyanates is that these materials are non-lubricating and also, by acting as solvents, can cause damage to pump seals.

7.3.1 Low pressure systems

Gear pumps are normally used for polyols. In a typical gear pump the components are not only bolted together but in engineering terms they are dowelled together, on taper pins, so as to ensure accuracy in alignment. As the gears are drawn around they trap the liquid between the gear teeth and the body of the pump. On smaller pumps accuracies of an order of tenths of a thousandth of an inch are required. Seals for the pump are usually made from PTFE and rub against a carbon steel block.

In the case of the isocyanate, non-lubricating liquids are being metered. One type of pump has been found extremely useful for handling such difficult materials, namely the annular piston-type pump. The annulae on this pump are made from carbon steel and white metal is used for the other internal components. The beauty of this piston pump is that it has only four moving parts and there are no springs or valves which may get stuck. Because of its simplicity this has been a very successful pump.

It does, however, have one problem: if a delivery rate of more than 50 cm^3/rev is required then the pump becomes inconveniently large and complicated. If more than one of these pumps is used to overcome this it may lead to other disadvantages, due to the increased complexity of the system.

In these circumstances other pumps can be used, including combinations of gear pumps and hydraulic swash plate pumps. These pumps must be modified to handle isocyanates.

In the USA it has been the practice to use pumps without the high degree of accuracy that we find in the types of pumps so far mentioned. There, pumps of the internal gear type, used in conjunction with a flow meter, are used. This combination is necessary, as this pump has larger clearances than precision type gear pumps and the resultant slippage which occurs needs to be corrected. Many machines of this type are in successful commercial operation.

Figure 7.2 relates percentage output to pump speed (in rev/min) at various back pressures for an accurate gear type pump. In the top curve a liquid of 100 cP viscosity is being pumped against a pressure of 100 lbf in^{-2} (0·7 MN m^{-2}). At 5 rev/min the volumetric efficiency is just over 93% and as the speed increases to 40 rev/min the pump is working at 99% volumetric efficiency. These results indicate an efficient pump.

If the back pressure increases to 500 lbf in^{-2} or 3·5 MN m^{-2} (a situation which can occur quite easily as a result of blockage), then, at the same speed of 40 rev/min, the volumetric efficiency falls to

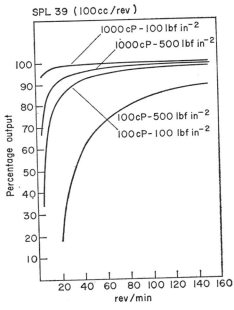

FIG. 7.2. Relation of output to pump speed. (Courtesy Slack & Parr Ltd., Kegworth, near Derby.)

96%. This is not acceptable, since a variation of ±1% is all that should be tolerated. What this means is that the whole of the pump and piping system would need to be examined and modified to ensure that we are not operating at the more critical pressure of 500 lbf in^{-2} (3·5 MN m^{-2}).

If we change the liquid to one with a viscosity of 100 cP then, at the same number of rev/min and at an operating pressure of 100 lbf in^{-2} (0·7 MN m^{-2}), the volumetric efficiency is only 92%. To obtain an efficiency of greater than 98% the speed must be increased to about 150 rev/min. The situation is much worse when the pressure increases to 500 lbf in^2 (3.5 MN m^{-2}), thus there is no point in using an expensive pump of this nature and one might as well use a cheap pump and correct its output subsequently.

Apart from having a good pump able to withstand the back pressure and poor lubrication conditions it is necessary to ensure that the pump is driven at constant speed. To this effect vee belts are not used in the drive system as they may slip. It is usual, therefore, to employ either a chain or a toothed-belt drive, which transfers the drive from an electric induction motor *via* a layshaft and flexible

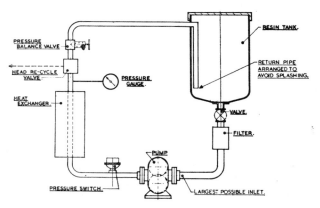

FIG. 7.3. Low pressure pumping system.

couplings. This arrangement ensures that no side loads are transferred to the pumps, which could interfere with accurate metering.

A schematic diagram of a low pressure system is shown in Fig. 7.3. The contents of the tank are capable of recirculation and included in the system are valves, a filter to protect the pump, a pressure switch which in the event of a blockage switches off the apparatus and a heat exchanger. The function of the heat exchanger is obvious. It might seem like gilding the lily but it must be remembered that reproducibility is the goal. This reproducibility can only come from accurate control, of which accurate temperature control is an important part. If the contents of the tank are recirculated and care is taken to discharge the liquid down the side of the tank, this would avoid air entrapment which would affect pumping efficiency. A pressure balance valve is also incorporated to balance the pressures between the feed to the mixing head and to the recycle.

PROCESSING EQUIPMENT FOR CELLULAR POLYURETHANES

7.3.2 High pressure systems

The high pressure metering unit uses diesel injector type pumps or hydraulic swash plate type pumps where the liquids are fed into the pump at a modest pressure, say 10–15 lbf in^{-2} (0·07–0·105 MN m^{-2}). They are then fed from these diesel injector or hydraulic swash plate type pumps to diesel injector nozzles (or to needle valves) where the pressures can be of the order of 1200–1500 atm. As the liquids enter the mixing chamber there is a pressure drop and they atomise to form small droplets and at the same time turbulence is promoted. The combination of turbulence and atomisation causes the particles to mix intimately and there is no further mixing required on these processes. These units are now being widely used in slabstock manufacture and also in integral skin foam technology (both rigid and semi-rigid).

In Chapter 6 mention is made of car bumper systems. These are manufactured on both low pressure and high pressure machinery at the present time, but there appears to be a growing trend towards the usage of high pressure systems for this application.

7.4 MIXING HEADS

7.4.1 Low pressure systems

These can be of the continuous or intermittent type. The continuous type may be used on slabstock machinery when it is normal to accept a small amount of waste at the beginning and at the end of the run. The size of the mixing chamber is large and a peg-type stirrer is employed, i.e. the central shaft of this stirrer carries a large number of pegs or projections.

Mixing heads of this type are used for blending liquids at approximately 500 kg (1100 lb) min^{-1}.

In the intermittent production of mouldings unacceptable blemishes can be produced if a small amount of out-of-ratio, or unmixed, ingredients are put into the mould at the beginning or end of the shot. Therefore, intermittent mixing heads have been devised in which recirculation valves are built into the head. The reactants can either be mixed by a high shear stirrer or recirculated back to their respective storage tanks.

The reactants are only in contact with the stirrer for about one-tenth of a second before they are discharged to the mould. If shots are being delivered every 20 sec then the self-cleaning action of the stirrer is usually sufficient to keep the head clean, but if a process is employed where the time between shots may be longer, e.g. 1–2 min, then the head will need to be cleaned by introducing solvent, followed by compressed air. A selection of high shear stirrers which fit inside appropriate barrels is shown in Fig. 7.4. These have a spiral cut on the outside of the stirrer basket to exert a pumping action on

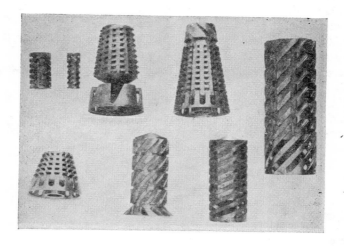

FIG. 7.4. High shear stirrers.

the liquids, which assists the stirrer to self-clean. To obtain the required degree of mixing these spirals are interrupted by vertical flutes. The largest of these stirrers is about 9 in long and is used for mixing approximately 100 kg (220 lb) of reactants per minute.

7.4.2 High pressure systems

Figure 7.5 shows a simplified diagram of a high pressure mixing head where the reactants are fed through diesel injector nozzles or needle valves. The sharp pressure drop causes the liquid streams to be finely atomised and these small droplets mix freely in the mixing chamber (which may be as small as 1 cm^3 in capacity). In addition, dissolved gases in the liquid (normally the isocyanate) are released by this pressure drop and these provide nucleating points around which

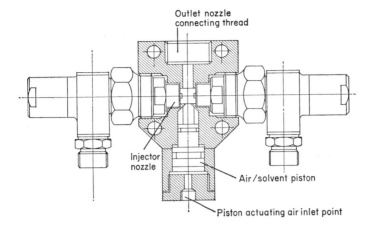
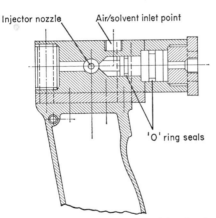

Fig. 7.5. High pressure mixing head.

the resulting foams form. If necessary the head may be cleaned by introducing first solvent and then a blast of compressed air.

Combinations of low pressure and high pressure heads can be used, but mainly in slabstock foam production.

7.5 MOULDS AND CONVEYING SYSTEMS

The dispensed mixed liquids can then be fed into various types of moulds.

7.5.1 Slabstock systems

These are highly automated plants that can be of up to 1000 lb (450 kg) min^{-1} capacity and have facilities for automatic width control. The formulation, output and colour can be changed without the plant closing down. Over 1m tons of flexible foam per annum is made, using techniques of this type.

The output from the reciprocating mixing head (Fig. 7.6) may be dispensed into a trough formed of paper. The trough is taken away at an angle so as to prevent the reactants underrunning the mixing head.

7.5.2 Foam moulding

Figure 7.7 shows part of a hot moulding line. These lines normally include provision for conditioning, automatic mould filling, automatic lid closing, an oven for the curing of risen foam, automatic mould opening, mould cleaning, and then precoating with mould release agent, and the cycle is again repeated.

The moulds are usually made of aluminium and this system differs from the slabstock plant previously described in that an external heat source is necessary, as there is insufficient exothermic heat generated within the foam to cause a satisfactory level of cure.

Multi-component metering units may be used in conjunction with such lines. It may be desirable to change formulation, output and pattern of laydown and this may be effected using punch card techniques.

Cold cure or high resilient foams are based on similar technology but diphenylmethane diisocyanate (MDI) or toluene diisocyanate/diphenylmethane diisocyanate (TDI/MDI) mixtures are used and the exothermic reaction reduces the amount of externally applied heat that needs to be supplied for curing in order to achieve the optimum physical properties. In order to meet automotive specifications, some additional heating is often necessary.

7.5.3 Rigid sandwich panels

A continuous rigid foam laminating machine is illustrated (Fig. 7.8). In this process the foam reactants are distributed between two substrates and a continuous sandwich of rigid foam is produced, usually in the range between 0·25 in and 6 in (6–152 mm) thick. A development of this machine is the inverse laminator in which the foam is sprayed on to a flexible facing which travels around a semi-

PROCESSING EQUIPMENT FOR CELLULAR POLYURETHANES 105

FIG. 7.6. High output slabstock plant mixing head.

FIG. 7.7. High output flexible foam moulding line.

cylindrical platen where the temperature is carefully controlled. The risen foam surface is then brought into contact with a lower facing and forms the eventual laminate. Commercial production work has been made with metal, asbestos, plasterboard, chipboard, hardboard and other building materials as substrates.

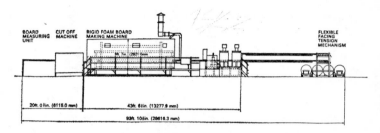

Fig. 7.8. Continuous rigid board machine (Courtesy I.C.I. Organics Division, Hexagon House, Blackley, Manchester.)

7.6 RECENT DEVELOPMENTS

7.6.1 Flexible foam slabstock

7.6.1.1 *Square block system*
In conventional slabstock manufacture the foam has a domed top which is largely due to the frictional drag effect of the side wall paper on the rising foam. A commercially successful method of reducing this effect has been patented and results in savings of more than 4–5% over conventional foam processes. This is done by introducing rising polythene sheets between the sides of the paper and the foam so that there is in effect a rising side wall which grows with the foam, thus reducing the friction.

7.6.1.2 *The Foamax process*
A more recent development is the Foamax process, illustrated in Fig. 7.9. In this case the metered and mixed liquids are fed into a fixed trough where partial expansion takes place and eventually the partially expanded material falls over a weir down an adjustable, inclined plane. The top surface of the foam is maintained at a constant horizontal level. The foam therefore expands vertically downwards and the frictional effect is thus reduced.

This process has now been in commercial operation for about 18 months and is showing an advantage in material saving of about 4–5% over the square block process, or between 8–10% over conventional domed top blocks. As the current world market for slabstock is about 1·3m tonnes such processes could result in savings of 100 000 tonnes of valuable raw materials. Investigations are being carried out in order that the process can be adapted for 'on the fly' variable width production.

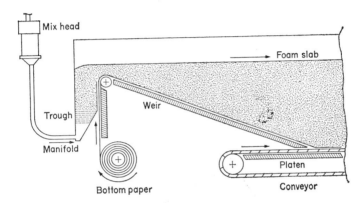

FIG. 7.9. Foamax process. (Courtesy Unifoam A.G., Zurich, Switzerland.)

7.6.2 Moulding

Recent developments include 'in the skin' moulding. An example of this is the Skin Form process in which a cover, made of PVC or a textile material, is placed over the mould. Vacuum is applied and the foam reactants are introduced. The heat of reaction, combined with the vacuum on the skin and the pressure developed, thermoforms the skin. This process normally produces cold cure flexible foams of high resilience and with skins of high definition.

7.6.3 Automatic polyurethane pipe manufacture

As a replacement for moulded pipe sections, a new process, Hexalag, has been developed. Figure 7.10 shows the principle of this process, in which a steel mandrel is spirally wound with a layer of paper.

A second paper is coated with polyurethane foam reactants from a traversing mixing head and the coated paper is then wrapped around the paper-covered mandrel. Alternative materials, such as aluminium foil and asbestos paper, have also been used. The wrapped pipe is hauled off with eight caterpillar-type tracks that need to rotate with the pipe.

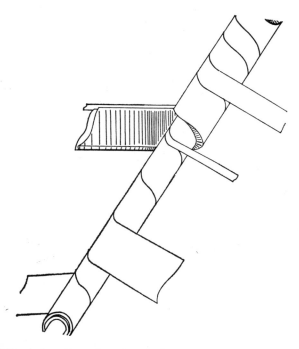

Fig. 7.10. Automatic polyurethane pipe manufacturing process. (Courtesy I.C.I. Organics Division, Hexagon House, Blackley, Manchester.)

7.7 THE FUTURE

This paper has only touched on the many processing opportunities of urethanes, but continuous development in both chemicals and machines is taking place to ensure that materials are safely processed and used and to obtain the maximum benefits of the versatility of urethanes.

There are still many problems to be solved, but the inventiveness and the skill of the industry will ensure that the full gains from urethanes can be utilised.

CHAPTER 8

Powder Processing of Thermosetting Plastics

R. H. CHAMBERS
(*Telcon Plastics Ltd*)

8.1 INTRODUCTION

The onset of plastics powder coating was the discovery, in 1943, of a means of producing low density polythene powder by a precipitation process. The product could be used to produce coatings by preheating an object and immersing it in the static powder. The powder melted and adhered to the article and was subsequently smoothed out by further heating in an oven.

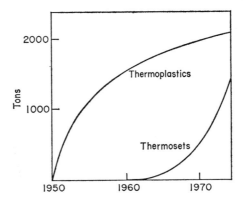

FIG. 8.1. Powder consumption in the UK.

This method was too crude to excite much interest and it was not until 1953, with the introduction of the fluidised bed from Germany, that there was commercial exploitation on any scale. Shortly afterwards, other thermoplastics including PVC and nylon were introduced and in the late 1950s epoxies were also being developed as coating materials.

The 1950s also saw the introduction of electrostatic techniques, initially for the efficient distribution of particulate fertiliser, and the two technologies had combined by the 1960s in the electrostatic deposition of fine epoxide powders to produce thin coating films.

The growth pattern shows the early dominance of thermoplastic materials which were used mainly in fluidised beds compared with the later rise in the use of thermosets which are extensively applied by the electrostatic process (Fig. 8.1).

8.2 BASIC REQUIREMENTS FOR A SATISFACTORY THERMOSETTING POWDER

In order to achieve a suitable coating powder, certain requirements need to be satisfied. Obviously, in the first place, the ability to reduce the material to a suitable size, both conveniently and economically, is of prime importance. In this respect the thermosets score over thermoplastics, for much less energy is required to reduce a brittle uncured material. In contrast, thermoplastics often require either freezing to below the brittle point, or precipitation techniques, before very fine powders can be obtained—neither of which process is very convenient.

A second important requirement is for the constituents of a system to be composed mostly of solid materials with softening points well above temperatures which can be reached in storage. Fine powders have a large surface area, which means that the presence of liquids, and temperature or pressure-sensitive constituents, increases the dangers of blocking.

The reactivity of a thermosetting system needs to be carefully controlled in order that during both compounding and subsequent storage premature curing does not take place. However, rapid curing at stoving temperatures around 160–220°C is very desirable.

During the stoving operation the viscosity of the melt must be sufficiently low to allow the material to flow into a smooth and homogeneous coated film prior to the onset of cross-linking, but should be high enough to prevent the coating pulling down from the edges and dripping from the substrate.

Finally, being in a finely divided state, powders are prone to adsorption of environmental contaminants. The choice of constituents is, therefore, limited, particularly to those having low sensitivity to atmospheric moisture.

8.3 COATING MATERIALS

By far the most important thermosetting materials used for powder coatings are the epoxides. Typically, the bisphenol A–epichlorhydrin resins with an epoxide equivalent of approximately 900 are used. These have melt viscosities sufficiently low to ensure an adequate flow out and, at the same time, softening points around 100°C which is high enough to prevent powder blocking. Improvements have been made to these resins to obtain better film properties by the control of molecular weight distribution and the elimination of gel particles. Another development has been the use of dimer acids to act as chain extenders to produce more flexible coatings.

The choice of cross-linking agents suitable for powder coating compositions is limited to those which are sensibly unreactive at the compounding stage yet can cure the coating rapidly on stoving. Originally, dicyandiamide [Fig. 8.2(a)] and BF_3-amine adducts [Fig. 8.2(b)] were used but these required stoving times of up to 30 min at 200°C and produced coatings with either poor surface finish or inferior mechanical strength.

The introduction of accelerated amine curing agents has allowed the manufacturer to produce powders of good surface quality which will cure in a few minutes at 200°C. These curing agents have revolutionised the thermosetting powder market and are the basic cause of the rapid growth in consumption since 1968. The use of these systems also gave the impetus for manufacturers to change from batch processes to continuous compounding.

The accelerated amines are extensively used in decorative coatings but for particular purposes other systems may be employed. Thus, for chemical resistance and heat resistance, acid anhydrides, such as pyromellitic dianhydride (PMDA) and trimellitic anhydride (TMA), may be used and aromatic amines (e.g. DDM—diamino diphenyl methane) have been employed for pipe coating where extremely short cure schedules are needed.

The outstanding adhesion, chemical resistance and excellent mechanical properties of epoxide coatings have resulted in their use in a wide variety of applications. However, the epoxides based on bisphenol A suffer from a tendency to yellow and chalk on outdoor exposure. To overcome these defects investigations into other epoxides, such as the cyclic aliphatic types, are proceeding but the most significant advances have been made into alternative systems, particularly polyesters and acrylics.

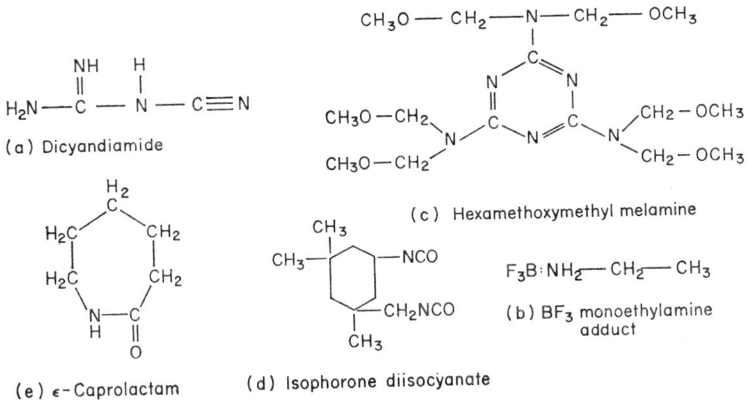

FIG. 8.2.

Polyesters, suitable for powder coating, are based on long chain polyols, cross-linked with either anhydrides, melamine derivatives or isocyanates. PMDA is the most common anhydride used and gives low gloss finishes with excellent heat resistance. A typical melamine derivative is hexamethoxymethyl melamine [Fig. 8.2(c)]. It produces coatings with good gloss, excellent resistance to ultraviolet radiation and good physical properties. It is unfortunate that the cross-linking reaction is by a condensation mechanism; hence with films greater than 100 μm (0·004 in) thickness the mass of material traps the condensation products, causing a porous structure. Undoubtedly the isocyanates produce the most attractive coatings, having high gloss and excellent mechanical properties coupled with excellent resistance to ultraviolet light. The isocyanate [Fig. 8.2(d)] is blocked to prevent curing at low temperatures and this is typically achieved with caprolactam [Fig. 8.2 (e)] which decouples on stoving and acts as a solvent, thus promoting flow out. Unfortunately, these systems tend to be extremely expensive and there is danger of powder blockage occurring, presumably due to premature release of caprolactam. In addition, at stoving temperatures caprolactam is volatile and unless care is exercised it will evaporate in the stoving oven and then condense on the cooler parts of the processing equipment.

Despite the wide use of acrylics in high quality paints, they have so far made little impact on the powder coating market. The basic problems appear to be high melt viscosity and low reactivity. However, a considerable amount of development work is taking place on these systems and recently it was announced in Japan that an acrylic polymer with pendant glycidyl groups, cross-linked with polycarboxylic compounds, was now available.

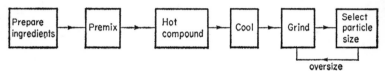

FIG. 8.3.

Although epoxides dominate the market at present and can be expected to be the single most important thermosetting powder in the foreseeable future, it may be anticipated that polyesters and acrylics will eventually extend the use of powders into areas where good outdoor weathering resistance is important.

8.4 MANUFACTURE OF THERMOSETTING POWDERS

The manufacturing process may be represented by a block diagram (Fig. 8.3). It may be necessary to reduce ingredients to a suitable size for premixing and subsequent feeding to the compounding stage. It is being realised that efficient premixing is an important step towards achieving consistent and homogeneous products and to this end high-speed mixers of the Henschel type are used to ensure sufficient dispersion of the ingredients.

The compound stage was originally performed by heating the material in a Z-blade mixer for about 30 min and then pouring out into open trays for subsequent cooling (Fig. 8.4). This process was satisfactory for slow curing systems and had the advantages that the individual components could be added when required and that control of colour was easier. However, batch to batch variation was frequently observed and the introduction of the faster curing systems

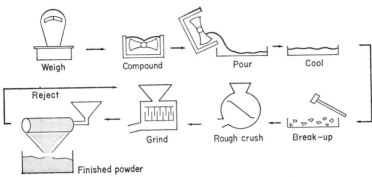

FIG. 8.4. Production of thermosetting powders—batch process.

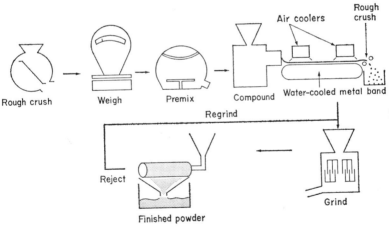

FIG. 8.5. Production of thermosetting powders—continuous process.

necessitated the use of continuous methods in which the compound is subjected to elevated temperatures for a few minutes only. Nowadays, the use of continuous methods is universal and plant manufacturers are offering complete lines capable of output measured in tons per hour (Fig. 8.5).

Because of the high reactivity of modern systems, hold-up in the compounding unit must be prevented at all costs and it is therefore necessary to use machines which act as positive pumps with 100% displacement of material. The normal single-screw extruder is unsatisfactory for the purpose but a wide variety of alternative equipment is available. These include twin-screw extruders, the Ko-Kneader, the planetary extruder and the Werner–Pfleiderer ZDSK compounder.

The extrudate is in the form of a thin strip between 1 and 3 mm (0·04 and 0·12 in) thick which is fed on to a water-cooled metal belt

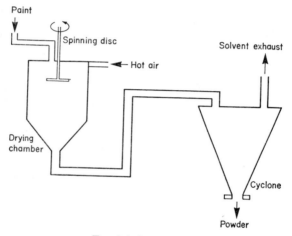

Fig. 8.6. Spray drying.

with ancillary air blowing on top. The cooled product is brittle enough to be rough-crushed by water-cooled rolls into a suitable size for feeding to the grinding equipment. Grinding to a suitable size may be performed by means of pin mills, with subsequent sieving or by attrition using air classification. Oversize particles are fed back for further grinding.

An alternative means of producing powder by spray drying has been proposed (Fig. 8.6). In this process the ingredients are dissolved in a solvent using conventional paint equipment. The solution is then drip fed on to a spinning rotor or sprayed into a drying cabinet. The solvent is flashed off and the dry particles collected in a cyclone. Since tinting is undertaken in the wet stage, accurate colour matching is possible and the direct and accurate control of particle size distribution obviates the need for grinding and sieving. Although at first sight the process appears attractive, it involves a high capital cost and films of powder prepared by spray drying have low gloss and are highly textured.

8.5 COATING TECHNIQUES

As already mentioned, the fluidised bed was first used for powder coating and although of minor importance for the application of thermosetting powders nevertheless in certain specialised processes the technique is still used.

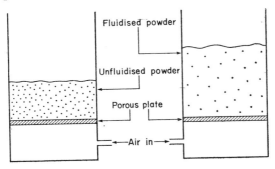

FIG. 8.7. The fluidised bed.

Essentially, the fluidised bed consists of two chambers separated by a porous plate (Fig. 8·7). The upper chamber contains the powder and the lower is a plenum chamber fed by an air supply. Fluidisation follows from the fact that powder will behave as a liquid if a gas (usually air) of sufficient velocity is forced upward through it so that each particle becomes buoyant in the ascending airstream. In the bed each particle is separate, one from the other, so there is little resistance to the passage of an object through it and the powder is able to flow around the article with comparative ease. Thus processing times are shortened and the coating of large and complex articles, with reasonable control over thickness, becomes a practicable proposition.

The minimum velocity of the gas through the bed is dependent on the particle size and density of the material. For plastics materials of density around unity, a particle size of between 50 and 250 μm

(0·002 and 0·010 in) is normal and an air volumetric flow rate of 5 ft^3 min^{-1} per square foot of bed area (equivalent to 1·5 m^3 min^{-1} per square metre) is adequate. The pressure below the porous plate should be sufficient to maintain this rate of air flow and it rarely exceeds 1 lbf in^{-2} (0·007 MN m^{-2}) above atmospheric.

For efficient fluidisation the powder should be free flowing. This is achieved by making the particle as near to spherical in shape as possible (this is because elongated particles tend to bind the powder together) and ensuring the absence of fine powder which also tends to promote blocking by virtue of its large surface area.

The article to be coated is heated above the melting point of the powder and is immersed into the bed, which is usually maintained at room temperature. The powder will melt and adhere on to the article. The thickness obtained will depend primarily on the time of immersion, the thermal capacity of the article and its temperature. As a general rule, agitation of the object in the bed is desirable in order to obtain a uniform coating.

The most important outlet for thermosetting powders using the fluidised bed technique is in the encapsulation of small electronic components such as resistors and capacitors. A series of machines has been developed by Messrs. Badalex which are capable of coating up to 35 000 items per hour on an automatic or semi-automatic basis. These machines consist of multiple preheating and coating stages, whereby the components are dipped in the bed up to a predetermined level. Multiple dipping is necessary in order to achieve thick coatings when items of low thermal capacity are being processed.

The advantages of fluidised bed coating can be quickly summarised: (*a*) the coating is completely solventless, thus eliminating fire hazards; (*b*) there is little contamination of the environment, which means the process can still be used where 'clean air' legislation is in force; (*c*) waste is virtually eliminated, as only the powder which adheres to the article is actually used; (*d*) high output rates on a mass production basis are readily obtainable but small-scale operations may readily be undertaken; (*e*) there is little danger of drips such as is encountered with paint and paste; (*f*) relatively thick finishes with good chemical and electrical resistance are quickly obtained.

Essentially, the fluidised bed is used for applying coatings greater than 200 μm (0·008 in) in thickness. For thin coatings other methods have to be employed, of which by far the most important is the use of electrostatic techniques.

In principle, electrostatic coating consists of feeding powder through the nozzle of a spray gun which is connected to a source of high potential (Fig. 8.8). The powder is charged and, on emission from the gun, is attracted to the nearest earthed object which is arranged to be the article for coating. The powder adheres to the article which is then transferred to an oven where the coating is fused into a coherent film.

There are two main charging mechanisms relevant to the electrostatic charging of powders: (*a*) corona charging; (*b*) triboelectric charging.

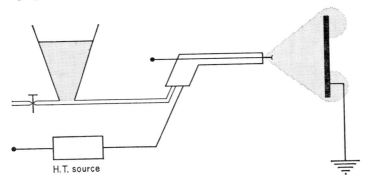

FIG. 8.8. Electrostatic powder coating.

Corona charging is almost exclusively used in the powder coating process. In this process air is subjected to an intense electrical field near an edge or point at high potential and forms either negative oxygen or positive nitrogen ions, depending on the polarity. The plastics powder is passed through the ionised field and acquires a charge, the value of which depends on the ion current density and the time spent in the charging region.

Triboelectric charging is caused by the frictional effects of powder particles rubbing against one another. The amount of charge required depends on the work function of both the substrate and the particles and is influenced by the conditions of the surfaces and the relative humidity. The spray gun does not require a generator, which should lead to considerable cost savings in equipment. A number of practical problems still have to be solved but it appears that tribo charging has the potential to be at least as effective as corona charging.

In order to retain its charge the powder is required to have a high electrical resistivity. This condition is satisfied with most plastics powders which have volume resistivities in the order of 10^{13} Ωm at room temperatures.

The powder, on leaving the gun, is attracted to the article to be coated at a velocity dependent on its charge. As it is normal to apply powders inside a spray cabinet with extraction facilities, it is important to ensure that the particle velocity is large compared to the air velocity within the cabinet. As charge is proportional to surface area, larger particles are more effective. However, there is an upper limit to the particle size when the gravitational force is of the same order as the electrostatic force. For plastics this means an upper limit of about 100 μm (0·004 in). In practice, a particle size range of 10–80 μm (0·0004–0·0032 in) is found to be effective, although undoubtedly a narrower range would be preferred if it could be economically achieved.

On striking the earthed substrate, the charged powder adheres to it by virtue of the electrostatic forces present. Powder will continue to be attracted to the article until the accumulated charge in the deposited layer is sufficient to repel further particles. In other words, there is a self-limiting effect. This occurs when materials of high electrical resistivity, such as plastics, are used but with powders of volume resistivity less than 10^8 Ωm, such self-limiting is not a feature and the powder layer is built up uniformly with time.

Although at first sight the self-limiting effect may appear to be of positive benefit, in practice this is not so. The thickness of the powder layer is not only of the order of 100–150 μm (0·004–0·006 in), which is greater than normally required, but it is also difficult to control being dependent on ambient conditions, in particular relative humidity. Further, it has been noted that under these conditions oncoming particles may still strike the powder surface and erode previously deposited powder, thus causing imperfections to appear in the final coating.

The efficiency of the process depends on a number of factors. The importance of high electrical resistivity has been stressed and materials possessing a volume resistivity in excess of 10^{13} Ωm are generally satisfactory. However, resistivity is sensitive to moisture and as changes of two orders of magnitude have been observed, with a change in relative humidity of about 30%, variations in coating performance could in practice arise from this cause.

Another variable is the potential applied to the gun. Most commercial apparatus is capable of applying up to 150 kV of either polarity. There is a great deal of controversy as to the conditions for optimum performance and as this will depend on the article and the powder being sprayed, no hard and fast rules can be made. In practice, settings of 50/70 kV at negative potential are fairly common.

Although electrostatic techniques are particularly suitable for applying thin plastics films, it is possible to achieve thick coatings up to and beyond 500 μm (0·02 in) by preheating the substrate. This causes the powder to melt on to the article, thus preventing subsequent dislocation. It may also be considered that the increase in temperature lowers the volume resistivity of the powder to below the level where the self-limiting effect operates.

The shape of the object to be coated influences the efficiency of the process markedly. Articles of large area such as flat panels are ideal, for the extent of overspray is small. In contrast, an open object such as wire work attracts only a proportion of the powder directed at it and up to 80% may end up as overspray. Here the advantages of powders become manifest, for it is perfectly feasible to recover oversprayed powder provided reasonable precautions are taken to prevent contamination. Most gun manufacturers offer collection cabinets as standard equipment which are capable of recovering over 95% of the overspray.

One of the features of coating wire work is the phenomenon known as 'wrapround', whereby the powder which passes the article is attracted to its rear surface. Thus, it is possible to coat both sides of such an object by spraying from one side only.

By the nature of the process it is difficult to electrostatically coat into recesses owing to the Faraday cage effect, whereby charged powder is attracted to the mouth of the recess but cannot penetrate into the recess. Methods of overcoming this have included the provision of an ancillary air supply to the gun which enables a jet of powder to be forced into the area. Recently, however, developments in gun design have led to the introduction of a system whereby the total power applied to the gun reduces as it approaches the substrate; this overcomes the Faraday cage effect and enables excellent penetration to be achieved.

A large number of production units coating a wide variety of articles has been installed. The process is basically similar to a normal paint coating line, as shown in Fig. 8.9.

Pretreatment may be by simple solvent degreasing or, where performance is important, by phosphating or shot blasting.

Coating is undertaken in a recovery booth using one or more guns, either manually or automatically. When required, some advanced spray equipment is available which is capable of automatically moving the gun in all three directions around a complex article. Coating thicknesses of the order of 50–75 μm (0·002–0·003 in) are normal, but there is a natural tendency to reduce this by careful control of coating conditions. Thus, thicknesses in the order of 25–35 μm (0·001–0·0014 in) have been claimed in the coating of flat panel work, although this is not typical at present.

The advantage of recovery has been rightly emphasised but it should not be overlooked that the economic success of the process relies on the ability of the user to prevent cross-contamination

FIG. 8.9.

between different powders occurring. This can be a serious problem when a small number of articles is being coated in a variety of colours, for there is no way of separating powders once they have been mixed.

Stoving may be carried out in recirculating hot air ovens or by means of infrared panel heating. As a general rule the latter is most satisfactory for flat panel work when shadowing and distance effects do not operate, but for general utility indirectly fired hot air ovens are to be preferred. Curing schedules depend on the size and shape of the article and typically vary from 5 min at 200°C to 15 min at 170°C.

The advantages of electrostatic powder coating over paint spraying may be summarised as follows: (*a*) the coating process is completely solvent free; (*b*) there is little contamination of the environment; (*c*) there is a high economy of usage, as overspray powder may be recovered and used again; (*d*) the process is well suited to both unskilled manual application and automatic application; (*e*) runs and sags are eliminated; (*f*) no undercoat or primer is necessary; (*g*) equipment energy costs are less; (*h*) one-coat powder finishes give protection previously obtained by multi-coat finishes; (*i*) smaller wrapping costs of finished articles are possible because of the superior abrasion resistance of powder coatings; (*j*) it is cleaner working with powder, rather than with paint.

Other means of applying powders to substrates have been developed which utilise the advantage of electrostatics.

One of these methods is the cloud chamber, in which a cloud of charged particles is introduced into the chamber and remains airborne by the application of a number of auxiliary air jets. The airstream causes the cloud to rotate within the chamber, thus causing any article introduced to be subjected to a continuous shower of charged particles. The density of the cloud can be varied depending on circumstances and considerable flexibility is possible by adjustment of the powder distribution heads, the position of the air inlets and the velocity of the air. It is possible not only to get one continuous cloud but also to vary its geometry in such a way that two cloud systems are produced in the same chamber, each rotating in opposite directions.

Powder particles which eventually leave the system drop to the bottom of the chamber where they may be recirculated with fresh material. This closed loop principle ensures the full use of powder without any additional complication of powder recovery.

The system, which establishes a slight negative pressure inside the chamber to prevent powder egress, is completely enclosed thus, reducing the problem of dust contamination.

Another technique which has found favour in certain applications is the electrostatic fluidised bed method, which combines the principles of both coating methods. In this process a number of charging electrodes are situated above the surface of the fluidising powder which causes a cloud of charged powder to form above the bed. There is a considerable vertical density gradient through the cloud which means that only about 15 cm (6 in) of working depth is available. However, the method is very suitable for coating such items as strip on a continuous basis.

Other methods of application have been developed which utilise the principle of spraying powder on to a preheated object. Thus, simple flock spraying, flame spraying and even the plasma torch have been used with some success, although outlets are few. One area which has excited considerable interest for a number of years is the insulation of rotors and stators for fractional horse power (FHP) motors. The advantage of powder coating over conventional slot lining lies in better performance and the fact that a less bulky insulation leads to substantial economies in copper winding wire. In addition, the coating gives a new freedom in the design of motors not previously possible with conventional methods. Several techniques have been devised for applying the powder, including both the fluidised bed and spraying of powder into the slots of a previously heated stack. These methods, however, suffer the disadvantage that, in order to prevent powder reaching areas which have to remain uninsulated, it is necessary to resort to masking, and build-up of resin on these masks leads to delays in production. To overcome this problem a new method has been devised which, by applying the powder only to those areas to be insulated, eliminates the need for masking altogether, and consequently production rates in excess of 500 coated stators per hour can be achieved.

8.6 THE COMMERCIAL EXPLOITATION OF POWDER COATING

When epoxide powders were first introduced it was predicted that their excellent chemical resistance and electrical insulation properties would lead to exploitation in the area of technical coatings. In Europe this has proved to be incorrect, for it was soon realised that these materials presented an opportunity to produce thin coatings which had real advantages over conventional paint systems. Thus, the electrostatic application of powders enables tough decorative finishes to be obtained in one application without the attendant hazards and inconveniences associated with solvent-based paints. As a result, 80% of thermosetting powders are now being used in direct competition with paints, and increasing concern with atmospheric pollution coupled with shortage in supplies of many common solvents has accelerated this trend. Further, as the overspray is recovered almost all substrates can be automatically finished with powders, which overcomes problems associated with shortages of the skilled labour required to reduce overspray losses with conventional paints.

The price of powder coating is comparable with a two-coat paint system and this can be true even with single-coat finishes, when, as is not unusual, the spraying efficiency falls below 40%.

Since powder coating generally has lower labour costs and fewer rejects compared to paint spraying, it will become even more attractive in the future as an alternative to stoving enamels. The extent to which this is realised would depend to a large degree on whether the single most important section of the finishing industry, the automobile manufacturers, change over to powder coating. Already the coating of car wheels is a well-established powder application and a great deal of work is being undertaken in both Europe and the USA in the use of powders for priming and top coat finishing of car bodies. The biggest problem associated with the use of powders for topcoats is the requirement for rapid changing of colour. Equipment manufacturers have been very active in designing production units which simplify colour changes and currently on the Continent some companies producing bicycles are satisfactorily spraying up to ten different colours. However, the requirements of the automobile manufacturers are much more exacting and at present the solution to the problem is not apparent, although several pilot plants have been installed and experimentally coated vehicles have been successfully produced. The use of powders for priming is much further advanced. Of particular interest is the combination of powder coating with electrophoresis to produce a superior priming surface which is then finished with a single topcoat of conventional paint. The coating of under-the-body members is also well established, a well-known example in this country being the powder coating of the Leyland bus chassis to give outstanding corrosion resistance.

Although in Europe decorative finishing has been the main outlet for thermosetting powders, the reverse is true in the USA, where the use of powders has been concentrated in the technical field. Of particular note is the use of epoxide powders for the external corrosion protection of large diameter pipe for gas and oil transmission. An outstanding example was the use of epoxide powder to coat 800 miles of 48 in dia. (1300 km of 1·2 m dia.) pipe for the Alaskan oil fields. This single application consumed over 1000 tons of powder and necessitated the construction of special coating units to apply the powder on site.

To forecast the future is always a hazardous undertaking, particularly in a rapidly developing situation where problems of supply of both raw materials and energy further complicate the situation. At present the consumption of thermosetting powders in the UK is about 1000 tons per annum, which represents only 0·5% of the industrial paint market. The prediction has been made that this will eventually reach 20% and, although this does not appear unreasonable, how quickly it will be achieved is an open question.

Certainly if past performance in other countries is any guide, powder coating has a bright future and is destined to become an important sector of the surface-coating industry by the end of the decade.

CHAPTER 9

FRP Low Pressure Processes

R. W. K. COOK
(*Fibreglass Ltd*)

9.1 INTRODUCTION

Before describing the various low pressure moulding processes available to the FRP industry at large, it would be beneficial to those who may not be familiar with the glass fibre raw materials to have a description of the basic fibre producing technology.

The fabricator has a basic objective of marrying together the performance opportunities offered by glass fibres and laminating resins to the processes available. Some understanding of the basic technology may lead to a better selection of the raw materials for any particular application being considered.

Glass fibres are basic to the family of structural plastics materials which utilise a very wide range of thermoplastic and thermosetting resins. The incorporation of glass fibre into resins changes them from relatively low strength, brittle substances, into strong, resilient structural materials.

In many ways, fibreglass reinforced plastics can be compared to reinforced concrete, with the glass fibre performing the same function as the steel reinforcement and the resin acting as the concrete. In the case of reinforced plastics, the glass fibre has high strength and high modulus and the resin has low strength and low modulus. The resin, despite its low strength and modulus, has the important task of transferring the stress from fibre to fibre, so enabling the glass fibre to develop its full strength.

9.2 BASIC GLASS TECHNOLOGY

9.2.1 Composition of some common fiberisable glasses

In order to understand many of the basic properties of glass and glass fibres and their beneficial effects in reinforcing plastics, it is best to examine the structure of the glass reinforcing material itself. Both the bulk structure and the surface structure of the glass fibres are important in determining the physical properties and the weathering properties of the glass reinforced plastics.

The major constituent of most commercial inorganic glasses is silica. The silicon dioxide molecule is composed of a central silicon ion surrounded by four oxygen ions in a tetrahedral configuration. In quartz, the room temperature modification of silicon dioxide, the tetrahedral molecules form an ordered three-dimensional structure. When melting takes place, the ordered network breaks down and becomes random in nature. As long as the silica is in the molten stage, the bonds in this more or less random network are constantly breaking and reforming. Upon cooling, however, the molten glass becomes too viscous to allow the silica molecules to revert readily to an ordered crystalline arrangement. The final structure can be visualised as a three-dimensional amorphous polymer. Molten silica can incorporate almost every chemical element. For example, sodium ions or calcium ions would fit into the 'holes' or interstices within the random network. These secondary ions or 'network modifiers' break some of the oxygen–silicon bonds and this makes the glass more fluid when molten. Under certain conditions various added ions (Al^{3+}, Mg^{2+}, Zn^{2+}) may even enter actively into the silicon network.

The chemical composition and common physical properties of glasses which are commercially used to produce fibres are given in Table 9.1.

'E' glass, with its low alkali content, has exceptional resistance to water attack. On the other hand 'C' glass, with a higher SiO_2 content than 'E' glass, has good resistance to acid attack but only moderate resistance to water attack. 'S' glass is a high modulus high tensile strength glass. 'A' glass is used for non-critical applications. This product has now been eliminated from commercial production in the UK. 'R' glass is a comparatively new glass fibre exhibiting high strength properties near to those offered by 'S' glass.[2]

TABLE 9.1

COMMERCIAL GLASS COMPOSITIONS SUITABLE FOR FIBRE PRODUCING AND THEIR COMMON PHYSICAL PROPERTIES[1,2]

Property	Type of glass				
	'R'	'E'	'C'	'S'	'A'
Composition (%)					
SiO_2	60	52·4	63·6	64	72·5
Al_2O_3 / Fe_2O_3	25	14·4	4·0	26	1·5
CaO / MgO	15	21·8	16·6	10	12·5
B_2O_3		10·6	6·7	—	Nil
Na_2O / K_2O		0·8	9·1	—	13·5
Specific gravity of fibres	2·55	2·54	2·49	2·49	2·45
Tensile strength of freshly drawn undamaged fibres					
$MN\ m^{-2}$	4 340	3 650	3 650	4 830	3 240
($lbf\ in^{-2}$)	(626 000)	(527 000)	(527 000)	(697 000)	(469 000)
Young's modulus of Elasticity					
$MN\ m^{-2}$	83 000	76 000	69 000	90 000	69 000
($lbf\ in^{-2}$)	($1·2 \times 10^7$)	($1·1 \times 10^7$)	($1·0 \times 10^7$)	($1·3 \times 10^7$)	($1·0 \times 10^7$)

TABLE 9.2

PROPERTIES OF COMMERCIAL 'E' GLASS FIBRES

	Glass fibres
Tensile strength, $MN\ m^{-2}$ ($lbf\ in^{-2}$)	2 480 (356 000)
Tensile modulus, $MN\ m^{-2}$ ($lbf\ in^{-2}$)	76 000 ($1·1 \times 10^7$)
Elongation, %	3·5
Specific gravity	2·54
Coefficient of thermal expansion, per °C	$4·85 \times 10^{-6}$
Heat distortion temperature, °C	Over 540
Creep	Does not creep

Details of the melting process and subsequent fibre producing technology are beyond the scope of this paper but it should be noted that very large capital investment has been put into the industry over the last 20–25 years. Over the years, glass fibre producing plants have continually advanced in technology.

'E' glass is the most common form of glass fibre used as a reinforcement and some common physical properties determined on commercially drawn filament are given in Table 9.2.

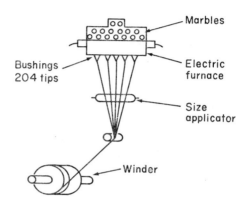

FIG. 9.1. A typical fibre-forming arrangement—marble bushing process.

It is of interest to refer to Fig. 9.1, which shows a diagrammatic representation of the fibre-forming process. After fibre producing, the filaments are coated with a chemical treatment, usually referred to as a forming size. The filaments are then drawn together over a shoe to form a strand, which is wound on a removable sleeve on a high speed winding head.

9.3 BONDING OF THE GLASS FIBRE TO THE POLYMER MATRIX

Coincident with the improved fibre producing technology described and more relevant to the subject of this Conference is the forming size. It is the ability of this size to adhere itself to the glass fibre surface and to link to the resin system which forms the basis of fibreglass reinforced plastics. A glass fibre surface is relatively smooth and polymers do not normally adhere readily to it. Consequently, stress transfer is normally low and difficult to achieve. Clearly, if significant use is to be made of the high strength of the glass fibre, the adhesion of the polymeric matrix to the glass fibre must be of a high order. Many properties do not require this high adhesion. Rigidity, thermal expansion and heat resistance are not significantly affected by adhesion. However, tensile strength and creep characteristics are significantly affected and it is here that the role of specific linking agents becomes extremely important, particularly in wet and humid environments.

9.3.1 Methods of application of linking agents and forming sizes

It is generally not appreciated by users of reinforced plastics that the great strength enhancement and resistance to fracture which occurs when one combines the high strength of glass fibres with a low strength, low modulus polymer matrix depends in the main on how well the polymer and the reinforcement are bonded together. The importance of the interface can be demonstrated by a simple calculation to determine the amount of interfacial area present in a typical laminate. For example, let us consider a plate of glass reinforced plastics, which is 10 cm (4 in) square and 1 mm (0·04 in) thick, consisting of 65% of glass by volume. Assuming the fibres are 13 μm (0·0004 in) in diameter, then the interfacial area in this plate is 450 000 cm^2 (6975 in^2) in 10 cm^3 (0·61 in^3) volume, and the entire area must be adequately bonded if optimum performance is to be obtained.

Most of the linking agents used today are either monomeric silicon compounds or chromium complexes and these chemicals have the ability to bond organic polymers to inorganic materials. It is believed that the function of the low strength polymer matrix is to transfer stress to the glass fibre reinforcement which then carries essentially the entire load. Since stress is transferred in shear through the interface, good adhesion of the polymer matrix to the reinforcement is essential. In the presence of water, some of the reinforcing action of the glass is lost due to water weakening or destroying the bond required to transmit the stress from matrix polymer to glass. Silane linking agents have demonstrated the ability to produce a bond of good initial strength (as shown by the failure of the composite by polymer rupture) and good retention of strength after severe wet ageing. Results shown in Table 9.3 show the general effect that can be achieved by the use of a silane linking agent.

When glass fibre is used as either mat, roving or chopped strands, it is not practical to use solely a silane linking agent on the glass but a more complex mixture, usually referred to as a forming size, is applied to the glass shortly after the fibres are drawn. This performs several functions, including the application of the linking agent to the glass, the protection of the glass from inter-filament abrasion,

TABLE 9.3

FLEXURAL STRENGTHS OF ROD STOCK MADE FROM SILANE SIZED 'E' GLASS STRANDS AND POLYESTER RESIN

Type of surface treatment on glass	% Glass content by weight	Dry	Flexural Strength 2 h Boil in water	72 h Boil in water	Units
Without silane	70	1 100 157 000	550 79 700 (48·4)	225 32 700 (79·5)	MN m^{-2} lbf in^{-2}
With silane	67	1 310 185 000	970 139 000 (25·0)	330 48 400 (70·0)	MN m^{-2} lbf in^{-2}

Figures in brackets refer to per cent degradation in flexural strength.

as a processing aid during manufacture. It thus optimises performance in the end use. Typical forming sizes consist of:

(1) A *linking agent* which can typically be an organosilicon compound or a chromium complex.
(2) A *film-former*, which can be from a wide variety of polymers, usually in emulsion form. Examples of commonly used film-formers are polyvinyl acetate latices, starch and emulsified polyester or epoxide resins which are usually applied in the form of an aqueous emulsion. The film-former protects the glass filaments during manufacture and subsequent handling and gives some integrity to the bundles of individual glass filaments. It also gives a stable forming package, usually referred to as a cake, which is later converted into the finished product, e.g. mat, roving or chopped strands.

(3) A *lubricant* such as certain silicones and acid amides can be used to lubricate filaments whilst they are being drawn, and aid passage through and over guide points, etc.

(4) *Anti-static agents*—various organic and inorganic compounds can be used to reduce the build up of static.

In the mid-1940s, Goebel and Iler[3] filed a patent on the use of methacrylato chromic chloride as a size ingredient and this was sold under the trade name Volan, later Volan A. A year later, Steinman[4] illustrated its worth by formulating a size with a film-former such as polyvinyl acetate, applied as an aqueous emulsion. This formulation is the basis for the chrome size still popular for rovings, where clean chopping with the minimum of filamentisation is required.

The surface of glass fibres contains many silanol groups after exposure to moist air and this surface absorption can only be removed by heating to 500°C in high vacuum. The methacrylato chromic chloride can chemically absorb on the surface of the glass by initial hydrolysis followed by dehydration to form Si—O—Cr linkages at the surface. It could also be chemically absorbed at all the negatively charged sites, since these are strongly acidic at pH of 3–6, which was the range originally used for the finish. The double bond of the complex is now available to react with vinyl type double bonds in the polymer by a free-radical mechanism (Fig. 9.2).

This methacrylato chromic chloride complex was the first finishing agent used which had the function of bonding both to the glass surface and the polymer matrix. Since that time many silane coupling agents have been developed and depend for their effect on this same principle.

A glass surface, a typical silane and a glass fibre surface are given in Fig. 9.3.

9.3.2 Recommended linking agents for different polymer matrices

At one end of the linking agent molecule are reactive organic groups which are capable of entering into typical organic reactions, and organofunctional silanes with these reactivities can combine with organic polymers both of the thermoset and thermoplastic type. Among the more common reactive organic groups are vinyl, alkylamine, methacryloxy, alkyl, mercapto and epoxy radicals. The

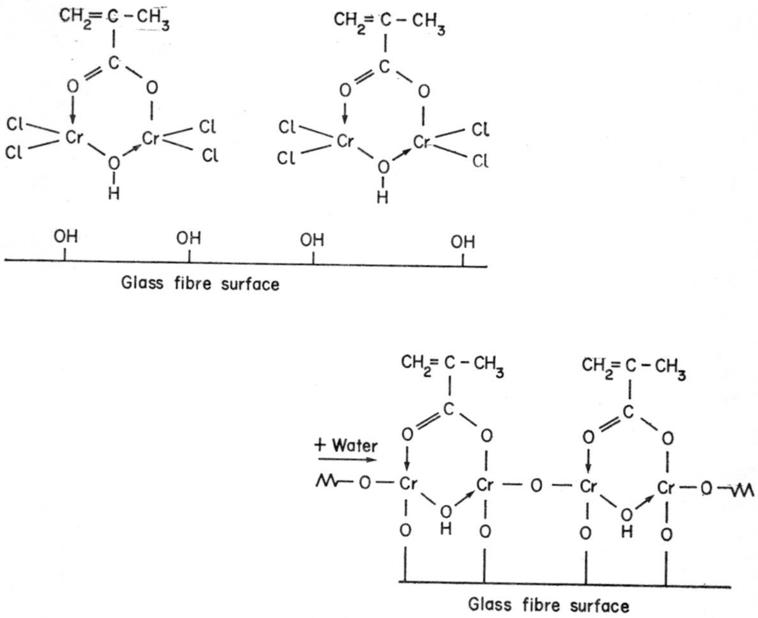

FIG. 9.2. Diagrammatic representation of the reaction between methacrylato chromic chloride and a glass fibre surface.

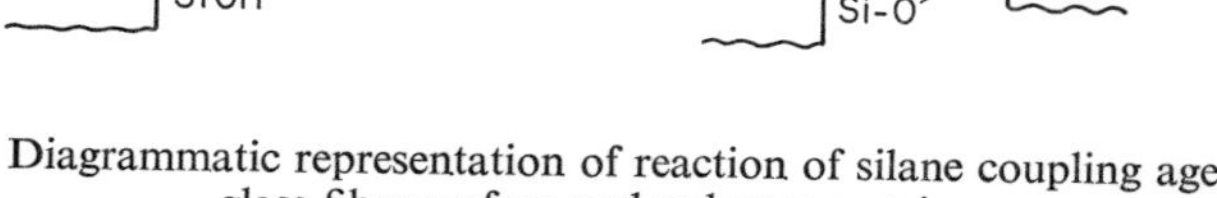

FIG. 9.3. Diagrammatic representation of reaction of silane coupling agent with glass fibre surface and polymer matrix.

chemical types of a number of silane monomer compositions used for thermoset polymer systems are shown in Table 9.4.[5]

The linking agent molecule also contains hydrolysable groups, typically alkoxy or halogen. Theoretically, one hydrolysable group

'UNION CARBIDE' SILANE LINKING AGENTS RECOMMENDED FOR VARIOUS POLYMER SYSTEMS BASED ON RESULTS USING HEAT CLEANED GLASS CLOTH

Recommended Union Carbide silane	Chemical nomenclature	Chemical composition	Thermosetting polymer
A-150	Vinyltrichlorosilane	$CH_2=CH-SiCl_3$	Polyester
A-151	Vinyltriethoxysilane	$CH_2=CH-Si-(OC_2H_5)_3$	Polyester
A-172	Vinyl-tris(β-methoxyethoxy)-silane	$CH_2=CH-Si-(OCH_2CH_2OCH_3)_3$	Polyester
A-174	γ-Methacryloxypropyltrimethoxysilane	$CH_2=C-C-CH_2CH_2CH_2Si-(OCH_3)_3$ $\quad\;\;\; \mid \;\; \parallel$ $\quad\;\; CH_3 \;\; O$	Polyester
A-186	β-(3,4-Epoxycyclohexyl)ethyltrimethoxy-silane	(epoxycyclohexyl)$CH_2CH_2Si-(OCH_3)_3$	Polyester, Epoxide
A-187	γ-Glycidoxypropyltrimethoxysilane	$CH_2-CH-OCH_2CH_2CH_2Si-(OCH_3)_3$ $\;\;\;\backslash\;O\;/$	Polyester, Epoxide, Melamine, Phenolic
A-1100	γ-Aminopropyltriethoxysilane	$H_2N-CH_2CH_2CH_2-Si(OC_2H_5)_3$	Epoxide, Melamine, Phenolic
A-1111	N-bis(β-Hydroxyethyl)-γ-aminopropyltriethoxysilane	$(HO\,CH_2CH_2)_2N(CH_2)_3Si(OC_2H_5)_3$	Epoxide
A-1120	N-β-(Aminoethyl)-γ-aminopropyltrimethoxysilane	$H_2NCH_2CH_2NH(CH_2)_3Si(OCH_3)_3$	Epoxide, Phenolic

per silicon atom would be sufficient if the coupling agent could be applied under anhydrous conditions. Water, however, is the preferred application solvent. The use of silanes in water results in the hydrolysis of the silane to a silanol, which is the active species in bonding to glass. Competitive with the condensation of a coupling agent silanol, with silanol on the glass surface, is a condensation of silanols on two coupling agent molecules. Therefore, the use of a silane coupling agent having one or two hydrolysable groups per silicon atom can result in the formation of di- or polysiloxanes having little ability to bond to glass under the normal use conditions. For these reasons and because of the solubility requirement for aqueous application, all commercially used coupling agents have three hydrolysable groups per silicon atom.

The organic group on the silicon that is hydrolysed to form the silanol has little effect on the performance of the coupling agent in the resulting composite. It is therefore chosen on the basis of ease of handling, nature of hydrolysis by-product (hydrochloric acid, acetic acid and alcohols) and, occasionally, by ease of synthesis.

Much of the research to date on silanes has been done using woven glass cloth as the reinforcement, with a silane finish. This was an easy way of evaluating the effectiveness of different silanes as finishes. (Commercially, a complete formulation has to be developed, consisting of a coupling agent, film-former, lubricant and anti-static agent, etc.)

The heat-cleaned fabric was used as a control and all silanes that were evaluated were applied as finishes to a 181–112 style fabric. The finishing solutions were normally aqueous, although some of the silanes, due to their solubility characteristics, were applied from a water–alcohol finishing bath. The treated fabric was allowed to air dry and applied at a loading of 0·5 wt %, based on the weight of the fabric. In those instances where a peroxide was used, it was applied to the silane-finished glass cloth from a toluene solution and the toluene removed by evaporation at room temperature. The mechanism of the linking agent and polymer matrix interaction is not completely understood. The literature contains apparently conflicting views on the types of silane that function best with specific types of polymer and also reports failure of different workers to obtain the same results when a given silane is evaluated in presumably the same polymer system.

9.4 LOW PRESSURE PROCESSES

9.4.1 Introduction

If one can apply the word classical to a very young industry such as glass reinforced plastics manufacture, then the classical method of low pressure processes is typified by the contact moulding or the hand lay-up process. This, and other more mechanised techniques of combining together glass fibres and thermosetting resins, are described.

Under the very open definition of low pressure processes, almost anything applies in the glass fibre moulding industry. The only process technique which may be considered outside the definition must be matched die moulding.

9.4.2 Commercial reinforcements

Commercial reinforcement is supplied in a number of forms which are designed for use by the fabricator of reinforced plastics, such as chopped strand mats, rovings and bi-directional materials such as woven rovings and glass fabrics.

Chopped strand mat (CSM)
This glass fibre product consists of chopped strands, nominally 5 cm (2 in) long, which are deposited in a random manner in the form of a mat. The strands are held together by a resinous binder. Binder types are generally based on polyvinyl acetate and polyester resin which exhibit good moulding characteristics and compatibility with polyester resins. CSMs offer excellent interlaminar cohesion and impact strength. Furthermore, when used alternately with woven rovings, they give adequate all-round mechanical strength for most composite laminate applications.

Continuous strand mat (Cont.SM)
This form of mat consists of a random lay of continuous multifilament strands. A small percentage of binder of low solubility is used to hold the strands together.

Continuous roving
This consists of several strands (each containing hundreds of filaments) gathered into bundles without twisting. They are then wound

into a cylindrical package. The package is designed to protect the roving during transportation so that the continuous roving can later be unwound easily for further processing. Packages can be unwound from the inside or outside; a common package size is 15·9 kg (35 lb) weight. The number of strands (or ends) per roving determines the linear density. Most products are in the range of 2400–4800 tex. The fewer the strands within a roving, the lower the tex (one tex is one gramme per kilometre of fibre length).

Woven roving

As the name suggests, this is a woven glass form using rovings as warp and weft and is available in various forms suitable for particular applications. Laminates produced from woven rovings exhibit high tensile strengths but poor interlaminar cohesion and, as indicated, are used with CSM to produce the necessary composite laminate structures.

9.4.3 Hand Lay-up[6]

This technique is one of the oldest in the reinforced plastics industry. Major advantages are that a minimum of equipment is required and the mould can be made at relatively low cost, using wood, plaster, sheet metal or even reinforced plastics. Hand lay-up, under most conditions, only gives one good surface, the working surface. Before any laminating takes place, the mould is prepared with a release agent to ensure that the moulding does not adhere to the mould. This is generally followed by a resin rich layer, known as a gel coat. The gel coat gives a surface with maximum resistance to corrosion. Frequently, this resin layer is lightly reinforced with a glass tissue or with an inorganic fabric or tissue. Once the gel coat has been allowed to partially harden, further resin is applied and the bulk of the CSM reinforcement is then built up alternately with application of resin. After the application of each layer of reinforcement, it is thoroughly impregnated with resin, and special rollers are used to consolidate the material and to help remove any entrapped air. In general, curing takes place at room temperature, but heat is sometimes used to speed the curing.

Besides the advantages of minimum equipment and low cost tooling, the contact moulding technique also has the benefit of minimal restriction of size, and relatively easy design changes. The disadvantages are that the labour content is quite high and the quality of moulding depends to a very large extent on the skill of the operators. The laying-up of the moulding and cure are both comparatively slow, although quite high production rates can be achieved under the right conditions.

Applications for this technique have the widest range, from small industrial mouldings through to large ocean-going yachts and fishing boats.

9.4.4 Spray-up[7]

Instead of using sheet materials impregnated with resin, i.e. chopped strands mats or fabrics, a production method utilising the simultaneous deposition of polyester resin and glass—spray-up—was one of the first to mechanise the use of glass fibre. Roving is chopped into approximately 5 cm (2 in) strands and projected with catalysed resin on to the mould. Rolling and consolidation of the laminate takes place as in the hand lay-up method. Advantages offered by this approach are an increase in throughput of laminate for a comparatively low capital investment in equipment and the use of the cheapest form of glass fibre reinforcement. A disadvantage is said to be that of controlling laminate thickness.

One particular advance in the use of glass fibre can be highlighted in the manufacture of tanks using a modified spray unit. This unit is mounted inside a rotating mould on a hydraulic ram and fed by a catalyst injection system with a combined resin and catalyst pump. The spray-up unit deposits resin–glass matrix on to the inside surface of the rotating mould. The centrifugal force of the rotating mould and special spring-loaded rollers contribute to consolidation of the laminate. A final layer of resin can be added as the resistant inner working surface.

This technique is said to be very competitive with conventional filament winding in many structures, due to the random orientation of the glass fibres. The glass content is around 30–35% by weight and hence savings are offered in raw material cost over equivalent conventional filament-wound structures which have glass contents of at least 60%.

9.4.5 Pressure bag moulding

The processes described in the previous two sections do not use a moulding pressure. Techniques do, however, exist which utilise some pressure and thus give an improved laminate (Fig. 9.4).

The tailored bag (normally rubber sheeting), is placed against the lay-up and fixed to the mould lips, while a strong lid is secured to the mould. The space between the bag and the lid is put under air or steam pressure, normally up to 3·5 atm. A modification to this method is autoclave moulding. In the autoclave an air or steam pressure of 3·5–7 atm can be reached.

Advantages include higher glass fibre loadings and improved laminate properties (due to reduction in number of voids) whilst undercuts are possible when required by the design of the laminate moulding.

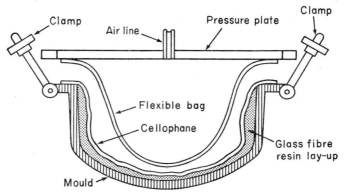

Fig. 9.4. Pressure bag moulding. A tailored bag is placed against the lay-up. Air or steam pressure is applied between the bag and plate.

A further variation of the above is that of vacuum bag moulding (Fig. 9.5), where a flexible tailored bag (synthetic rubber, polyvinyl alcohol film or polythene) is used.

The bag is placed over the lay-up, as with the previous method, and clamped down around the mould circumference. Air is now evacuated from the space between the mould and the bag. The resultant pressure (up to atmospheric pressure) consolidates the reinforcement, forcing out entrapped air and surplus resin.

In general, the laminate surface in contact with the bag is improved over that of previous processes.

9.4.6 Filament winding[8]

Possibly the major advances in using glass fibres have been in the process of filament winding. This process has played an increasing part in the construction of large storage tanks and chemical plant pipework.

The technique consists of winding continuous glass strands on to a mandrel. The mandrel normally rotates whilst at the same time the roving feed and guide unit reciprocates along the length of the mandrel. This allows the whole process to be programmed precisely to position the rovings' bandwidth to predetermined patterns. Resin can be applied directly on to the mandrel but more generally by allowing the glass rovings to pass through a resin bath just prior to winding.

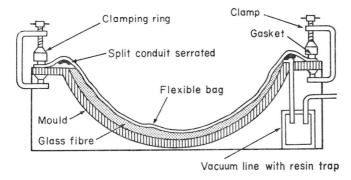

FIG. 9.5. The vacuum bag moulding method. Atmospheric pressure is applied to the resin impregnated lay-up by exhausting air from beneath an impervious flexible sheet sealed to the edge of the mould. The flexible sheet is laid over the lay-up and clamped round the mould. Vacuum is drawn and entrapped air and excess resin are forced out.

Applications which can use this type of process have the capability of utilising glass fibres to their maximum mechanical advantage. The products have the highest strength to weight ratio of any thermoset moulding. In such glass resin mouldings the bore may be made very smooth by the prior use on to the mandrel of any particular resin.

The resin content at the working surface can be increased and is then usually reinforced either by a glass or thermoplastic fibre tissue. Normal winding process can then be completed to the determined design parameters.

A number of companies specialising in the manufacture of large chemical fluid storage tanks have adapted the filament winding process so that winding slit-width, chopped strand mat and woven rovings may be used. This combination allows complete design flexibility in relation to the use of glass fibres in the production of large simple shapes.

In the USA some companies specialising in the production of chemical process piping have refined the winding process to such a degree that the continuous high speed production of 5 cm (2 in) diameter epoxide resin glass pipe can be made competitive with steel pipe production.

Further pioneer work in high speed pipe production for low pressure applications uses a centrifugal casting process. Information available indicates the use of rovings applied to the inside of a rotating mould. The resin filler system is applied to the inside *via* pressure nozzles. Production rates, although not published, are said to be faster than filament winding. Use of rovings in casting techniques is a breakthrough, as previously only mats had been the reinforcement.

A specialised filament winding technique using rovings has been developed by the Kroyer organisation. It produces spherical tanks with the obvious advantages of largest capacity in relation to surface area. The tanks combine great strength, large capacity and low production costs. An additional advantage of the spherical shape is that it lends itself to automation and hence gives a very efficient production process.

Such tanks have been used for the storage of domestic heating oil but other applications for storage of chemical products and corrosive substances have been reported.

It should be recorded that a significant machinery development in continuous filament winding and take-off is that of the Drostholm process.[9,10] Drostholm have rationalised the mass production of filament wound tanks into standardised sizes and they process pipe in continuous lengths. Diameters up to 3·5 m (11·5 ft) at a production rate up to 400 kg h^{-1} (8819 lb h^{-1}) can be achieved.

A specific example of a throughput rate is 22 m h^{-1} (72 ft h^{-1}) for 1·2 m (3·9 ft) pipe with a 6 mm (0·24 in) wall thickness. This is dependent upon the resin, catalyst and accelerator used.

This process utilises a mandrel of cantilever construction, with a surface moving continuously at a rate of 40 mm rev^{-1} (1·6 in rev^{-1}) in a longitudinal direction. The mandrel is automatically wound with a cellophane strip to protect the surface and to facilitate the eventual release of the pipe. A resin-rich inner surface is applied, using tissue or slitwidth mats impregnated with resin. A combination of circumferential and longitudinal disposed glass fibre rovings are impregnated and consolidated by a system of rollers. The hoop fibres are applied continuously whilst the longitudinal component is added by chopping roving to the desired length and applying to the mandrel.

Curing is carried out by a 5-stage oven heated by infra-red units. The whole operation is controlled by a central console allowing time to gel, cure and postcure as the tube progresses through the oven. Cutting to length is similarly controlled and is done by an automatic diamond-saw unit.

The improvement in physical properties in filament wound components is made possible by the accurate alignments of the reinforcing fibres in the best direction to meet the stress conditions. Optimum utilisation of materials is thus possible.

9.4.7 Cold press moulding

Cold press moulding is a production process suitable for reinforced polyester mouldings of up to about $5 m^2$ ($54 ft^2$) in surface area. In speed and economy it falls between the slower contact moulding technique and the more rapid hot press moulding process. Heated tools are not required and a pressure within the range 0.04–$0.17 MN m^{-2}$ (6–$25 lbf in^{-2}$) can often be used; elaborate press equipment is not essential and tooling costs are lower than in hot press moulding. Matched moulds are used, made either of reinforced plastics or steel. The exothermic heat of the curing reaction heats the moulds or dies and consequently the mould cycle shortens as moulding proceeds. When an equilibrium temperature is reached, the moulding cycle settles at its minimum, between 5 and 25 min depending on the size and thickness of the moulding.

Cold press moulding has proved to be economic for runs of between 130 and 8000 mouldings, depending on the complexity of the component and on the production time available. Reinforced polyester components produced by cold press moulding are smooth on both sides; they can have a reinforcement content up to 50%, as against 35% for contact moulded laminates, and are therefore of higher mechanical strength. The average moulding cycle is 12 min, compared with 2 h for contact moulding and 3 min for hot press moulding. Cold press mouldings are also usually more consistent in quality than contact mouldings.

Very large components, e.g. one-piece boat hulls or lorry cabs, can often be produced economically by cold press moulding.

For many applications, improvised press equipment is sufficient. This can consist of an angle iron frame and a hydraulic or manual device for raising and lowering the upper mould. For small components the frame can be dispensed with and two suitably constructed moulds are all that is needed, provided they are sufficiently weighted to exert the necessary pressure. For large cold press mouldings, the use of a hydraulic press is recommended. Accurate platen guides are always essential for good results and consistent quality. These can be incorporated in even very simple tools without great difficulty.

The success of good cold press moulding depends to a large extent on the design and construction of the moulds. Large mouldings and long runs are best made with steel dies, although these need not be as heavy and strong as those used in conventional hot press moulding processes, since the pressures involved are much lower. Reinforced polyesters or epoxide shell moulds, backed with a tough concrete or a resin/aggregate mixture, have proved satisfactory for many applications and such moulds can produce about 2000 good mouldings before needing attention. Reinforced plastics shell moulds are made by contact moulding over a pattern constructed from any convenient rigid material. The female (concave) mould is usually made first, and when this is fully cured, wax, equal to the thickness of the moulding, is laid up in it and the male mould is contact moulded on this.

9.4.8 Resin injection process[11]

The technique of resin injection as a moulding process is almost as old as hand lay-up. The aircraft industry in the late 1940s used to make laminates *via* this process as it gave a method of more precise glass content control.

With the advent of continuous strand mat reinforcement and more specialised equipment available to the industry, the resin injection process has received new emphasis.

It is a low pressure process using matched moulds (which can be made from glass fibre) the cavities of which are filled with reinforcement. The mould edges are then clamped which holds the reinforcement in place. The resin, catalysed and accelerated, is injected to a maximum working pressure of $0 \cdot 45$ MN m^{-2} (65 lbf in^{-2}). Air is pushed out to escape through the mould cavity *via* the flash line. When the mould is fully charged a small amount of resin seeps through the weakest point along the flash line. The injection point is quickly sealed with a plug to prevent air entering the mould whilst the resin cures.

Venting is fast and thorough. This stops air being trapped in the laminate. Mouldings are produced with accurate control of thickness with good tooling construction. The resin/glass ratio can be varied from 2:1 to 5:1. Smooth, glossy surfaces can be imparted to one side or both sides of the laminate by spraying or brushing gel coat on to the surface of the mould before the reinforcement is placed in position. The resin can be pigmented as desired.

When the resin has cured, it takes only a few minutes to remove the mouldings and repeat the process.

Mouldings with surface areas of up to $2 \cdot 8$ m^2 (30 ft^2) have been produced, but it is suggested that laminates up to $37 \cdot 2$ m^2 (400 ft^2) are said to be feasible.

Processes described previously, such as hand lay-up and spray lay-up, require skill and concentration by an operator to prevent air from being trapped within the laminate. Thickness and resin/glass ratios are difficult to control. Resin flow problems can also leave voids in cold press mouldings but the resin injection process is an advance over these techniques as air is pushed out by the advancing resin front. Accurate control of laminate thickness and resin/glass ratio are achieved using matched tools.

9.5 MECHANICAL PERFORMANCE DATA OF LAMINATES

Quoting mechanical performance data for glass fibre laminates can only be generalised, especially with reference to some of the processes described. Much depends upon the skill of individuals when using hand lay-up and spray lay-up techniques.

Data have not been specifically quoted for the pressure bag and vacuum bag moulding methods, but as a general rule they tend to give laminate performance some 10–15% above that quoted for hand lay-up for the same glass content by weight.

The specialised area of filament winding can have a whole range of mechanical performance data. This depends on angles of wind of the reinforcement roving, thickness of the laminate, etc., but it should be noted that the highest levels of mechanical properties are available from this technique.

9.5.1 Hand lay-up laminates

Typical measured data are of the following order. In an attempt to give some form of recognised standardisation of laminate property, all laminates were prepared as specified in BS 3496 : 1962, Appendix D with a glass content range of 39–42% by weight.

Tensile strength (BS 2782, Method 301L, 2·5 cm (1 in) width),
  150–165 MN m^{-2} (22 000–24 000 lbf in^{-2})

Flexural strength (BS 2782, Method 204B),
  270–285 MN m^{-2} (39 000–41 000) lbf in^{-2})

Flexural modulus (BS 2782, Method 302D),
  0·75–0·92 × 10^4 MN m^{-2} (1·09 × 10^6—1·33 × 10^6 lbf in^{-2})

9.5.2 Spray-up laminates

Results are quoted for a typical laminate made by the spray-up process. The laminate was made by movement of the resin/glass depositor in two directions at 90° to each other. The methods of determination of the mechanical properties of the laminate were as given above.

Tensile strength	74–84 MN m^{-2} (11 000–12 000 lbf in^{-2})
Flexural strength	120–128 MN m^{-2} (17 000–19 000 lbf in^{-2})
Flexural modulus	0·63–0·72 MN m^{-2} (0·91–1·04 × 10^6 lbf in^{-2})
Glass contents	24%

9.6 SUMMARY

Criticism has been levelled at glass fibre producers for keeping information about their processes and products relatively secret. In presenting this paper in this way it is hoped that potential users, specifiers and fabricators may be able to better understand the role of using glass fibres.

The mechanism of how glass fibres 'bond' themselves to resins may not be exactly understood, but there is no doubt that the surface chemistry applied at the fibre-forming stage is responsible for the glass reinforced plastics industry. The diagrammatic representation presented should allow an understanding of how the mechanical properties of a two-component composite (glass fibres and resin) work.

Resin has an important role to play in binding the fibres together, distributing the applied load between the fibres, and protecting the fibres from their environment. The mechanical properties depend upon the glass fibre reinforcement.

A presenter's licence has been taken to cover a wide range of FRP low pressure processes. All techniques described are in use within the glass fibre moulding industry. Still the largest single consumer of glass fibre reinforcement within the UK is the hand lay-up process. Despite its lack of sophistication when compared with other more mechanical techniques, the industry is still dependent upon it. Reinforced plastics applications should be judged on their merits and many examples of very long service life, often in hostile environments, are available.

The other processes described are an attempt to add respectability to the classical technique and they tend to claim more than their share of publicity. This is certainly the case with respect to filament winding. Many symposia have been devoted to the applications for this technique, especially in the USA.

There is still a place for FRP low pressure moulding processes within the thermosetting moulding industry and they have a continuing role to play in the future. Advances in moulding techniques will be made with the emphasis on reducing labour content, but the 'classical' method should remain.

ACKNOWLEDGEMENTS

Particular acknowledgement is due to Dr. J. D. Crabtree and Mr D. Pickthall, and also to the staff of Redland Pipes Ltd., Aladna Gerlings Ltd (Holland) and J. Coudenhove (Austria).

REFERENCES

1. FRP Design Data Book, Fibreglass Limited, 1969.
2. 'R' glass, Societe du Verre Textile, French Pat. No. 1,435,073.
3. Goebel, G. T. and Iler, R. K. U.S. Pats. 2 544 666; 2 544 667; 2 544 668.
4. Steinman, R. U.S. Pat. 2 554 910.

5. Sterman, S. and Marsden, J. G. Theory and mechanism of silane coupling agents in thermosetting resin systems, Union Carbide Corporation, Silicones Division.
6. Chemical Engineering in Fibreglass Reinforced Plastics, Design Series No. 6.
7. Coudenhove, J. Kunstoffe–Maschinen Ges. M.B.H., Marc–Aurel–Strasse 4/9, P.O. Box 465, 10.10 Vienna, Austria.
8. Aladna Gerlings Machinery N.V., Postbox 6, Aalten, Holland.
9. Drostholm Products, Vedback, Denmark.
10. Redland Pipes, Plastics Division, Poole, Dorset.
11. Bouvier Newlove (UK) Ltd., North Elmham, Norfolk.

CHAPTER 10

Machinery for Thermoset Injection Moulding

R. A. BUTLER

(*Butler-Smith Associates Ltd*)

10.1 INTRODUCTION

Injection moulding of thermoset materials has lately become more feasible and attractive economically because of improvements which have been made in their processing characteristics. It was not unknown in the past for an injection moulding machine to need fairly constant attention when processing thermosets, with settings for pressure, temperature and so on requiring frequent alteration because of the inconsistency of the material being put through it. Some years ago, for instance, a thermoset machine being run under test conditions needed readjustment every few minutes because of variations in the contents of the hopper. This is probably an extreme case but perhaps not entirely untypical of thermoset injection moulding, even until quite recently.

Makers of thermoset materials have clearly put in a great deal of work to improve the processing characteristics of these materials and today thermosets are not much more difficult to injection mould than thermoplastics. This is a great advance and it may not yet have become widely realised and made its full and deserved impact on the moulding industry.

It is possible to draw a parallel with thermoplastics injection moulding and, although not entirely relevant, comparisons are made between the thermoplastics and thermosets types of processing.

Only some 10 years ago terms such as 'exotics' and 'difficult' were applied to materials which at that time were tricky to process. It seems hard to believe today that nylon, polycarbonate, acetal resin and even ABS were described in such terms.

Many moulders who would never have dreamed of handling these materials only a few years ago now process them without any trouble. From the standpoint of the machinery designer and manufacturer, the picture used to be one in which only those moulders with higher than average technological skill and supporting facilities tackled difficult materials, and those moulders who did not have a high level of expertise confined themselves to the easier materials, chiefly the various grades of polystyrene and the polyolefins.

Today the picture is one in which the less technologically skilled moulders are able to process what were formerly difficult materials because the job has been made easier for them. The wider use of what were difficult materials has come about because their processing has been made easier. It is not necessarily because the level of technological skill of the average moulder has markedly risen.

This is not by any means a contemptuous thing to say. This is how things should be. It is the job of materials suppliers and machinery manufacturers to improve their products so that they become easier to use. In fact I believe this to be the philosophy of good machine design. A designer has no right to expect the people who will use his machines to have to learn more than is necessary.

So if, as seems likely, thermosets follow the same course as the so-called difficult thermoplastics of a few years ago, we can expect to see them being injection moulded successfully by many moulders whose technology and skill may be described as average. Many of the new entrants to thermoset injection moulding will have only average knowledge and aptitude, whereas the typical thermoset injection-moulding shop today is above average in its technical standards.

Easier operation of injection moulding machinery can be equated with practicality. This paper describes how thermoset injection moulding machines can be designed to be as practical as possible, dealing with the various parts of the machine in turn. But first there is one other background aspect worth mentioning. A great deal of thermoset processing at the moment is by time-honoured, labour-intensive methods, for example multi-impression compression moulding. If, as expected, thermoset injection moulding is going to become more widely adopted it may lead to the gradual abandonment of these old methods in certain cases. The point is that these old methods can hardly be described as 'high' technology. Therefore the process which takes their place must not involve too great a technological leap forward. It would not succeed if it were. This is why thermoset injection moulding must be as practical and as easy as material and machinery manufacturers can make it, so that conversion from old to new methods can be as painless as possible.

Machine design should not be done in an ivory tower but according to the practical needs of the industry, that is with the objective of making the task in hand as uncomplicated as possible. Machinery manufacturers sometimes lose sight of this and add complication to their machines when a fundamental rethink would indicate a simple solution. For example, basic inadequacies in machine performance are sometimes compensated for, and disguised by, additional control equipment which would be unnecessary if the design was right in the first place.

Let us imagine for the purpose of this paper that a thermoset injection moulding machine is to be designed from scratch rather than being adapted from a design originally conceived for thermoplastics. In some respects there would be no significant divergence between the two types, but in other respects one might expect and find shortcomings where thermoset machines began life as thermoplastics designs and were adapted with greater or lesser success.

10.2 CLAMPING UNITS

The direct hydraulic-ram clamping system is almost certainly better for thermoset applications than a toggle system, or for that matter a 'lock and block' system. The chief reason is a strongly practical one and, to repeat an earlier point, practicality goes hand in hand with easy operation.

Unlike most thermoplastics, the use of thermoset materials incurs the risk of highly abrasive dust becoming deposited on working surfaces. This is very detrimental to both toggle and to lock and block systems, with their various moving parts and frictional contact surfaces (Fig. 10.1). Consequently there is a high wear liability which results in mould misalignment. A direct hydraulic-ram clamping system is far less prone to the effects of thermoset abrasive dust. There is normally a wiper ring around the ram at the front of the main cylinder and this removes any deposits of dust as the ram retracts.

There is no effective way of dustproofing a toggle or lock and block system and undue wear, because of the abrasive dust, is virtually unavoidable. The automatic lubrication system of a typical toggle machine may counter the effect of dust to an extent, but it is hardly working under the best conditions.

A clamping unit giving generous daylight in relation to platen size is specially advantageous for a thermoset machine. Heated platens in particular take up available daylight, in effect adding to mould height. One might have a machine with adequate platen area but too

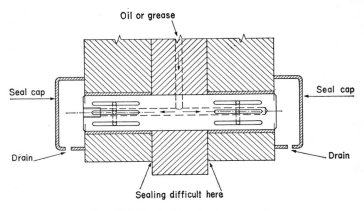

FIG. 10.1. Typical toggle pin assembly.

little daylight because of the requirement for heated platens, and in such a case the mould may have to be mounted in a machine which has the daylight required but with excess platen area and injection capacity. This may upset the whole profitability of the job. It is easier to design a clamping unit of the direct hydraulic-ram type to give good stroke and daylight in relation to platen size than it is with a toggle machine. This is because the length of stroke of a toggle machine is largely governed by its platen size, whereas there is no such constraint with a ram unit.

Another advantage of the direct hydraulic-ram system for thermoset moulding is that it consistently achieves its rated clamping force throughout the machine's life, whereas a toggle system may tend to give a diminishing amount of clamping force. This is due to the mechanical stretch and flexing of various components, in turn aggravated by wear. Any fall-off in clamping force is not always noticeable in thermoplastics moulding, as the surface skin of components hardens off almost instantaneously in the mould after injection and flashing does not occur readily. Thermosets, on the other hand, are in a fluid state in the mould for a comparatively long time, i.e. until they cure. If the clamping force should drop critically, and the material is still fluid, the mould will flash. This problem is made worse by the high penetrating power of thermoset materials in the fluid state.

Cycle times in thermoset moulding are generally much longer than those for thermoplastics and, with such long cycle times, there is less reason to attempt to provide a very high-speed clamping system. There is, however, a substantial advantage in having a rigid, accurate and robust unit capable of providing a good approach speed with controllable and sensitive sensing. The lower initial cost of the direct hydraulic unit is another substantial advantage.

Advantages might be claimed for the very fast types of clamping system but these are largely discounted by the fact that in thermoset moulding the clamping unit is motionless for a disproportionately large amount of the total time that the machine is in use. The opening and closing period is therefore a smaller proportion of the total cycle time.

This same fact also affects the efficiency of the automatic lubrication system of a typical toggle machine. This usually provides for a shot of lubricant each time the system is in motion. Under full clamping force there is a tendency for the film of lubricant between the frictional surfaces under load to be broken down or squeezed out. The longer that 'full clamp' is sustained, the greater this squeezing-out of lubricant from between the various components. Because long periods on full clamp are entailed with thermoset moulding, the effectiveness of an automatic lubrication system on a toggle machine is offset. The result must be a higher degree of wear. Also, under some conditions of moulding, the compacted material in a mould will expand as it sets and a higher clamping force than that applied will be generated. This adds further to the toggle-bearing loads and may even prevent mould parting.

Thermoset mouldings are more rigid, harder and less easily deformed than most thermoplastics and the moulds are frequently more expensive. Because of the tendency to flash a mould displaying any damage to its mating faces, there is a good argument for having a fast response from whatever mould sensing system is used.

If a lubricant is used on the tiebars of a thermoset machine, wipers are essential. Even so there is a tendency for the film of oil or grease to mix with the dust and produce a very effective grinding paste. For this reason a machine with dry lubrication, incorporating PTFE–bronze bushes for example, has an advantage. Hardened tiebars are a great advantage provided that sufficient care is taken to avoid notched shoulders.

It may be advantageous to have cast-in aluminium heaters, mounted on the platen, as a standard feature of a thermoset machine, each platen being a zone of temperature control with its own instrument. Other desirable features are provisions for platen cooling and the incorporation of an insulating material between platen and heater. A suitable material is Syndanyo, which is capable of taking a load of 2–4 tons in^{-2} (30–60 MN m^{-2}). A 0·3 in (7 mm) thick layer is normally sufficient.

Fully automatic working and insert moulding will often require central hydraulic ejection, and because of the tendency of thermoset mouldings to stick, a 'joggle' facility is valuable on the ejection system. This joggling is preferably done automatically during the run-out of the recycle timer. A limit switch interlocked to the return of the ejector ram is preferable to a pressure-switch signal which could give a false indication, for example if the ejectors stuck forward. A good system can be evolved by timing the forward stroke and returning to a limit switch which shows a safe condition for mould close.

An ejector comb, or sweep arm, which will knock the mouldings off the ejector pins is frequently useful although the joggle facility may be sufficient. If central hydraulic ejection is not fitted, a mould-halt facility is useful for insert loading. An air line close to the clamping unit for cleaning flash off moulds is also useful.

10.3 INJECTION UNITS

When considering injection units, we again find the desirability of different values for various parameters compared with a thermoplastics machine. In particular, shot volume, screw torque and injection pressure should be higher than for a thermoplastics machine of comparable size. For example, an injection pressure of

25–30 000 lbf in^{-2} (170–210 MN m^{-2}), may be needed, which is much higher than usual for thermoplastics work.

It is not possible to provide the screw of a thermoset machine with a check valve because of practical difficulties. As the material would have to flow through the small passages of a check valve, it could cure and harden in the valve and the practical problems involved with cleaning-out would be unacceptable. Good streamlining is in fact essential in thermoset screw design to minimise the liability to hold-up of the material flow.

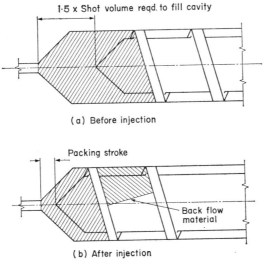

Fig. 10.2. Conditions before and after injection.

However, because there is no check valve, flowback of material over the screw end unavoidably occurs when the screw moves forward on injection, and this phenomenon presents special problems in thermoset processing (Fig. 10.2). In early thermoset injection-moulding machines, when no special measures were taken to counter flowback, as much as a quarter of the shot could be lost on injection. In other words, a quarter of the material which was ahead of the screw before it moved forward flowed back over it and only three-quarters went out through the nozzle. Even more material could flow back over the screw if a very long hold-on time was required and the screw would eventually 'bottom-out', making further application of pressure to the moulded material impossible.

Flowback presents special problems because the proportion of material which flows back on injection receives a second working by the screw and hence a further heat input. There may be a difference of 60°C between the hottest and coolest parts of the shot on this account. The ideal is to heat the material evenly. An obvious remedy is to prevent the screw from turning on injection. As the screw is forced into the material, one can imagine an Archimedean action of the material upon the screw, causing it to turn. On machines with

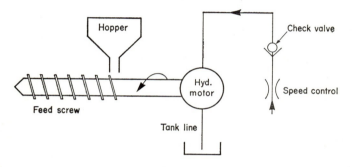

Fig. 10.3. Screw drive with inhibited reversal.

hydraulic screw drive, it is simple to provide a check valve in the hydraulic line connected to the screw-drive motor so as to prevent this (Fig. 10.3). On machines with electric or electromechanical screw drive, it is necessary to provide a brake of some description, which is more complex. Even so, some flowback of plastics material will still occur when screw turning is checked, and this is not a complete answer to the problem.

Flowback can be reduced if the screw has a fine-pitch flight. It is common for a screw to have a flight pitch equivalent to the diameter, but it is obvious that there will be less liability to flowback with a screw having a flight pitch of only 75% of the diameter, as a more obtuse angle is presented to the material by the flight as it moves forward.

There are other advantages to be gained from having a fine-pitch screw in combination with a direct high-torque drive. For example, a fine pitch gives a better conversion of available torque to thrust at the flight, permitting a shorter screw length for the same number of flights, and thus reducing the volume contained within the heated section of the plasticising unit and therefore the residence time of the material.

In an attempt to achieve a homogeneous shot temperature, some machinery manufacturers incorporate means to increase the rate of heat input to the material during screw rotation and hence compensate for the effect of flowback. At the beginning of screw rotation, the material which flows off the screw is that amount which flowed back on the previous injection stroke. This amount receives a secondary heat input. Once it has passed off the screw, it is followed by fresh material which has been worked by the screw just once and is therefore not as hot. Some machines, therefore, have programmed back pressure providing for an increase in screw back pressure during screw rotation. The first amount of material—that which has been heated twice—is processed at low back pressure, and the following fresh material is worked at higher back pressure. The higher the back pressure, the more the heat input into the material. It is seen that the object of this system is to produce a shot of homogeneous temperature and viscosity.

However, some of the advantage of this modification may be lost on injection. Much of the power required for injection is dissipated as friction as the material is forced through the nozzle. The nozzle builds up heat and in turn puts heat into the material flowing through it, with more heat going into the material at the end of the shot than at the beginning.

Another school of thought holds that it is not necessarily a disadvantage if the material at the back of the shot is cooler than that at the front. For one thing, this offsets the heating effect of the nozzle. Also, if the material at the back of the shot and against the screw end is at a lower temperature and therefore more viscous, it will have less tendency to flow back over the screw on injection.

In fact, there may be a case for *dropping* back pressure at the end of screw rotation to provide a high-viscosity plug or cushion for the end of the screw to bear against on injection. It may be possible to achieve such a degree of control that this plug is precisely the amount of the whole shot which flows back, and it can be brought up to the required temperature by the second working it receives.

In any case, screw back pressure should be controllable on a thermoset machine, not by regulating the discharge flow from the oil injection cylinder but rather by pressure regulation. This is necessary because of the high viscosity/temperature dependency of the material. Screw speed and the rate of injection may need to be controlled to closer limits of accuracy than tolerable with thermoplastics. Screw cooling may be desirable and air is preferable to water for this purpose as it is less severe and presents no leakage problems. However, screw design is the most relevant factor here. In the case of a deep-flighted screw (which in proportion to the

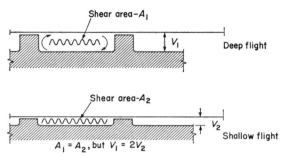

FIG. 10.4. Comparative shear work in screws.

volume contained in the flight has a small heated area available from the barrel), it has been proved that a heated screw offers advantages in improving the plasticising rate. In this case, any increase of screw speed would not have such a marked effect upon material temperature as in a shallow-flight screw, where the material is subjected to a greater amount of working (Fig. 10.4).

Accurate control of the temperature in the water or oil jackets around the barrel is essential, and there should be a copious flow of fluid through these jackets. The thermocouple for *control* of the fluid temperature should be in the barrel close to the screw. Better still, the temperature should be *read* close to the screw and *controlled* on the outer wall of the barrel. This will avoid hunting, incurred if control sensor and heat source are apart (Fig. 10.5).

If the jackets cover the front cap (i.e. the part which must be removed to allow removal of cured material) they must be easily removable without breaking water or oil lines. The front cap and/or the nozzle should be easy to remove and the female threads should be in the cap and not the barrel, so that spilled material on the internal threads may be cleaned out easily. A cooled nozzle is advantageous, as excessive heat may be picked up from contact with the sprue bush.

Screw delay may be useful but it can be employed only if the front cap is cool. If the front cap should get hot, the thin layer of the thermoset material left after the injection stroke will be cured inside it. If this is liable to happen, early screw rotation is called for as a precaution and screw delay is clearly inapplicable. The incidence of

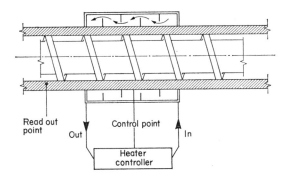

FIG. 10.5. Barrel heating layout.

unwanted curing in the front cap may depend upon whether heat is provided by the screw or the barrel heaters.

Screw flights need to be robust at the front in order to avoid damage when injecting against hard, cured residual material—a hazard which must be accepted by moulders and machine designers. This is why the width of flight on a thermoset screw is usually heavier (e.g. about 30% thicker) than on a thermoplastics screw. The heavy flight width will be consistent for the full length of the flight, right to the rear of the screw. In this case the channel volume of the screw is less than that of a thermoplastics screw of otherwise similar dimensions. If the flight width were gradually reduced from heavy at the front to moderate at the rear, there would be a compression ratio as the channel width is reduced from the back to the front of the screw. The designer has to consider the effect of this upon the material and the possibility of a premature cure in the barrel.

The hopper of the machine should be round so that a stirrer can be used if necessary for some materials, and it should be made of stainless steel. The hopper shut-off should be readily accessible and a lift-off facility is an advantage in order to assist in cleaning out residual dust.

An air line should be available for cleaning the injection unit and all rams and sliders provided with wipers. Again the enemy is abrasive dust, which may affect the control gear and other working parts of the machine. The hydraulic tank should be sealed against the entry of dust with a paper-element filter on the air intake. Similarly, the electrical cabinet should be dust-sealed and have a dustproof cover over the push-buttons and other switches. Totally sealed electric motors are ideal.

10.4 MOULDS

Regarding moulds, as much draught as possible should be allowed in component design, and in mould design the provision for efficient ejection should be carefully studied. Ejection is more difficult than with thermoplastics. The core and cavity should be in hardened steel, solid and with at least 0·003 in (0·07 mm) thickness of chrome for wear resistance. Sharp corners should be avoided if possible because of the abrasiveness of the injected material. Relieve all areas not associated with shut-off of the cavity or runners.

The mould should be designed to permit easy removal of flash. To minimise flash, there should be no bolt holes, dowel pins or slides close to the cavity unless there is an 'escape' groove between them (Fig. 10.6). Holes and slots for the escape of gas should be liberally provided wherever possible, especially as all ejector pins must have a very tight clearance.

Recently, many new techniques have been introduced into mould design, for example vacuum extraction of air may assist where gates have been badly placed either by lack of consideration or because of the dictates of component shape. Wherever possible, of course, one would try to make allowance for altering gate positions which could initially be wrongly placed.

Three-plate moulds are a convenient way of altering and indeed placing gates, and can lead to a reduction of waste, cured material.

Multiple gating of a moulding may assist filling and gas escape without additional runner length, and varied sizes of gate can alter flowpaths considerably.

Total elimination of waste runners can be achieved if a cold-runner system is used. This is the reverse of the hot-runner system in a thermoplastics mould, but the problem of streamlined flow is more difficult to overcome and generally the more sensitive and unstable materials are not useable in such a system. A slow cure is therefore demanded, with consequent lower production potential.

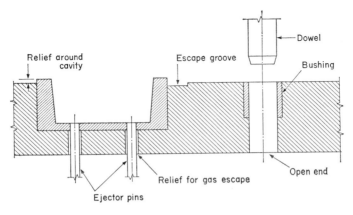

Fig. 10.6. Some points of mould design.

10.5 CONCLUSION

Finally, a word of advice to those about to enter thermoset injection moulding. Not only choose your machine with care, but choose your mould maker with care also. He may be good on thermoplastics moulds, but this does not mean that he knows about thermosets. A badly designed mould, or one with badly located and poorly matched heaters, will run badly in the best of machines, but it is not always possible for the moulder to tell who is to blame.

CHAPTER 11

Injection Moulding—Material and Process Requirements

F. J. PARKER

(*British Industrial Plastics Ltd*)

11.1 INTRODUCTION

Compression moulding has always been the principal process for the conversion of thermosetting plastics materials. The continual need for faster production, so as to reduce costs, led to the development of fully automatic compression moulding which reduced the labour costs of the process by automating the metering of material to the mould and removal of the completed parts. In addition, preheating techniques were applied to this development, so as to cut down the cure time necessary in the press. In these two ways, and by the development of faster curing moulding materials, the compression process has now been developed to such a fine art that it is unlikely that further significant gains in speed of operation, and thus cost savings, are possible.

Notwithstanding such development, the best rates of production on compression moulding are generally not as good as equivalent thermoplastic parts made by injection moulding and it was for this reason that attention was turned to the possibility of injection moulding thermosetting materials. This has now become a well-established process.

As with fully automatic compression moulding using preheating, the injection moulding process provides a means of metering materials to the mould cavities and preheating the material at the same time so as to reduce the curing time. However, the attraction of the injection machine is that it is capable of achieving both these processes so very much more efficiently, so that the overall costs for the product are significantly lower than compression moulding, despite the higher machine costs.

This paper is concerned with outlining the advantages and disadvantages of the injection process in both technical and economic terms, defining the requirements of the process and hence the characteristics required of the material, and discussing the optimum moulding conditions, which must include the specific requirements of proper mould design. Modifications of the basic process have been developed in the last few years, for example warm-runner moulds, injection–compression techniques and special methods for handling materials such as DMC. These are also described in this paper.

Although most thermosetting materials can be injection moulded, special injection moulding grades are normally necessary and raw materials suppliers have been very much concerned with developing such materials. At present the bulk of injection moulding thermoset materials is represented by phenolics, ureas and melamines and much of the discussion and illustration in this paper is centred around these materials. However, alkyds and diallyl phthalate materials (DAPs) are also injection moulded satisfactorily and, more recently, polyester resin based dough moulding compounds (DMC) have become possible materials for use in the injection moulding process.

11.2 ADVANTAGES AND DISADVANTAGES

Table 11.1 lists the qualitative advantages and disadvantages of injection moulding thermosets, compared with compression moulding, on the basis of part quality and cost.

The chief advantage, of course, is the very much faster rate of production which can be obtained with the injection process, especially with the slower curing thermosets such as phenolics, ureas and melamines. Generally, these can be moulded at three to four times the rate obtainable using the fastest compression methods. However, the inherently much faster rate of cure of the alkyds and DMCs usually only allows a gain on the injection process with these materials of about 20–30%. This faster production rate, together with the fully automatic character of the process, provides very substantial cost savings which are usually more than sufficient to offset those aspects of the process (for example, the higher machine cost and the wastage of sprues and runners) which affect the cost disadvantageously.

The cost saving is typified by the example given in Table 11.2, which shows production rates and costs for an electrical component produced by compression moulding in a 12-impression tool and by injection moulding in both 12- and 4-impression tools. These data

TABLE 11.1

ADVANTAGES AND DISADVANTAGES OF THERMOSET INJECTION MOULDING COMPARED WITH FULLY AUTOMATIC COMPRESSION MOULDING (WITH PREHEATING)

Feature	Quality	Cost
Fast cure rate		+
Efficient metering of material		+
Efficient preheating of material	+	+
Thin flash—easy finishing	+	+
Gate removal	−	−
No granular pattern on mouldings	+	
Possible flow lines on surface	−	
Wastage of sprues and runners		−
Possible stressing and distortion	−	
Press cost		−
Tooling cost (fewer impressions needed)		
Running cost and materials		−
Low labour costs		+

Plus (+) indicates that, in general, the injection moulding process may be considered advantageous in this respect; minus (−) indicates the reverse.

show a production rate on injection moulding of about three times that for compression moulding and, for the same number of impressions, a cost saving of about 21% accrues from the faster injection process. Alternatively, the faster rate of production on injection can be taken advantage of by reducing the number of impressions on the tool to four and thus saving on the initial capital outlay. In such an example, the cost saving is about 31% compared with the compression moulding example. Selection will very much depend on the number of components to be produced.

The same example is further illustrated in Fig. 11.1 which graphically illustrates (*a*) the equivalent cycle times and (*b*) the press times for equal output.

PRODUCTION COST DATA FOR 13 A PLUG-BASE FROM U.F. (COMPARED WITH COMPRESSION MOULDING, THIS COSTING GAVE 31% AND 21% SAVINGS FOR THE 4- AND 12-IMPRESSION INJECTION TOOLS)

		Injection	Injection	Compression
Number of impressions		4	12	12
Lifts per hour		120	100	36
Mouldings per hour		480	1200	432
Nett weight	oz/100	81·3	81·3	81·3
Gross weight	oz/100	90	90	118·7
Excess	%	10	10	46
Material		Urea (INJX)	Urea (INJX)	Urea (INJX)
Material cost	pence/oz	0·58	0·58	0·58
Press size (Bipel)		155/50	260/88	150[a]
Mould cost	£	3800	9600	10000
Moulding material cost	pence	52·5	52·5	69·25
Moulding labour cost/100	pence	2·63	1·04	2·88
Press cost/100	pence	23·4	11·25	26·63
Finishing cost/100	pence	6·25	6·25	1·25
Inspection cost	pence	3·75	3·75	5·0
Packing labour cost/100	pence	0·46	0·46	0·46
Packing material cost/100	pence	1·67	1·67	1·67
Total cost/100	pence	90·66	76·92	107·14
Tool amortisation cost/100 (over 2 000 000 mouldings)	pence	18·75	47·92	50·00
Total direct cost/100	pence	109·41	124·84	157·14
% Saving of injection compared with 12-imp. compression		30·37	20·55	—

[a] With infrared preheating.

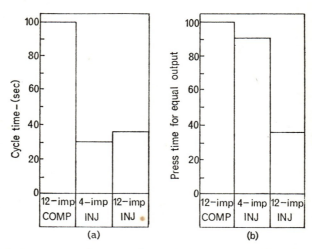

Fig. 11.1. Comparison of compression and injection production rates for 13A plug base in urea. (a) Cycle times, (b) press times for equal outputs.

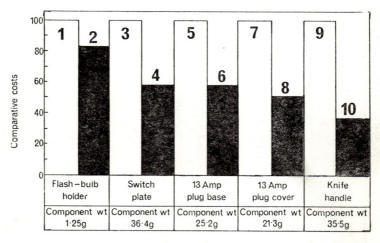

Fig. 11.2. Injection v. compression-moulding costs (urea-formaldehyde moulding material) for the five components. Tool costs are not included, 1, 3, 5, 7, 9: Compression. 2, 4, 6, 8, 10: Injection. 1, 2: 8-impression. 3–8: 2-impression. 9, 10: 1-impression.

The actual cost saving achieved by injection moulding is very dependent upon the moulding weight and thickness and, to a variable extent, the number of impressions in the mould. This is because of the relative economic importance of the sprue and runner system which has to be thrown away. Clearly, the cost of such scrap material can be economically very significant when making mouldings of small weight. On the other hand, with very much heavier mouldings the loss of the runner system can become relatively insignificant. Similarly, the gain in production rate, compared with compression moulding, is greatest for the thicker parts because it is with these that the efficiency of the preheating achieved on injection moulding becomes most evident. This is illustrated in Fig. 11.2, which gives a number of examples of mouldings (shown in Fig. 11.3) made from urea material.

The material cost also affects any considerations as a similar graph for the more expensive melamine shows (Fig. 11.4). With more costly materials, such as DAP, the economic advantage of the faster production rate of injection moulding can be completely negated by the wastage of material in sprues and runners. In such instances, compression moulding would be economically preferred unless the wastage can be avoided by the use of warm runner tools (this topic is discussed later in the paper).

The considerable gain in cure rate obtained when injection moulding phenolics, ureas and melamines is due not only to efficient preheating of the material in the barrel of the injection machine, but also to the considerable frictional heat built up as the material is pushed through the feed system and the gate. This usually means that the material enters the mould cavity at a temperature either equal to or above that of the mould itself.

The alkyd and DMC materials on the other hand do not benefit quite as much in this way. In part, this is because significantly lower barrel temperatures are used but also, more importantly, because the viscosity is so much lower that there is less frictional heat build up on passage through the feed system. Furthermore, the materials are inherently fast curing and therefore there is less potential gain possible for this reason. Thus, typical savings in cure time for DMC are usually only of the order of 25%.

However, overall production cycles for DMC are often substantially better than with compression, due to savings in press open time. This arises from the nature of the material which, being tacky

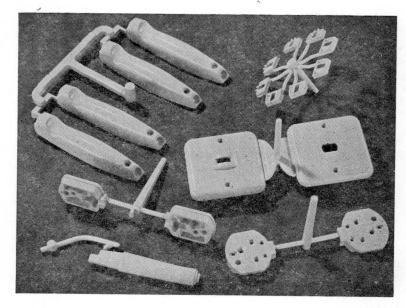

Fig. 11.3. Typical examples of mouldings made from urea-formaldehyde moulding materials.

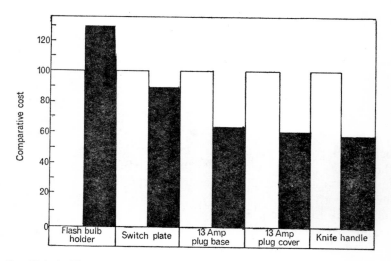

Fig. 11.4. As Fig. 11.2 but for more expensive melamine-formaldehyde moulding materials.

and fibrous, takes considerable time to weigh and dispense to the mould. In extreme examples, such as a multi-impression tool for a small part, the press open time can far exceed the cure time solely because of this difficulty. In such instances the production rate for injection moulding can be many times better than that for compression and then a similar economic gain accrues as when using some of the other thermosetting materials. It is therefore the fact that the injection machine provides a very convenient and efficient means of metering material that makes it most valuable for DMC.

Apart from the economic considerations of the injection moulding process, minor advantages also accrue because of the thinner flash obtained and because the granular pattern on the surface of mouldings, arising from the granular form of the moulding material, is avoided. However, generally speaking, the advantages of the injection moulding process are largely economic, whereas the disadvantages are mainly technical.

The disadvantages arise from two basic problems. First, it is possible to cause precure of the material before filling the mould and this, or the use of too high a pressure, leads to stressing in the moulding which may result in final cracking on ageing. Secondly, as the flow is very directional, flow marks and weld lines are likely to develop around or near to core pins where the flow is forced to divide. For these reasons, it is usual to check the quality of the moulding by stoving (at times and temperatures appropriate to the material), so that any stressing will be revealed by cracking or excessive distortion. Stoving for 48 h at 80°C is considered an adequate standard for urea and phenolic mouldings but many properly moulded parts will withstand up to 7 days at this temperature without serious degradation of the mouldings.

Distortion due to orientation can be a problem and this effect is usually associated with conditions where the shrinkage of the material varies considerably from one dimension to another in the moulding. However, this is usually avoidable as it is most often caused by the use of an unsuitable design or position for the gate. For example, Fig. 11.5 (b) shows how a straight tab gate on a thick section moulding allows the material to jet and snake into the cavity, leading to disadvantageous orientation and consequent weld lines. A side gate, shown in Fig. 11.5 (a), avoids this, the flow being distorted by impact with the opposite wall of the cavity.

Another example, Fig. 11.6, shows a switch plate moulding which, if fed with a small tab gate, as shown in Fig. 11.6 (a) tends to warp and show flow marks on the surface, where the flow is interrupted by the cored-out holes. The broader flash gate shown in Fig. 11.6 (b) eliminates the distortion problem and greatly reduces the unsightly flow marks on the surface.

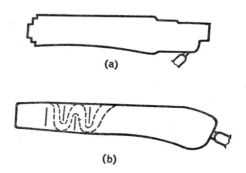

Fig. 11.5. Different tab gate positions on two knife-handle moulds: (a) side gating avoids cavity jetting and flow marks: (b) end gating produces a 'material worm' which leads to flow marks (shown dotted).

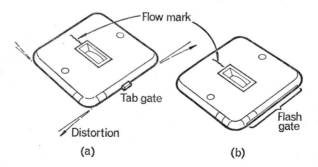

Fig. 11.6. A tab gate (a) caused the switch-plate moulding to warp badly in two directions. A flash gate (b) almost completely avoided this problem besides reducing weld lines or flow marks.

Inconsistent or high mould shrinkage is often considered to be an inevitable disadvantage of injection moulding but this problem can be avoided by use of the correct size and position of the gate. This is illustrated in Table 11.3 which shows the values obtained for shrinkage measured along two dimensions of a square plaque moulding. In one example the plaque was moulded with a small tab gate in the centre of one side and in the other example it was moulded with a flash gate extending almost the whole width of the side. Although both gates had very similar cross-sectional areas, the latter gave very consistent shrinkage values between each dimension, whereas the tab gate example had very variable shrinkage and, because of this, distorted very badly.

TABLE 11.3

COMPARISON OF SHRINKAGE OF TAB AND FLASH GATE FED PLAQUES

Moulding A: $\frac{1}{8}$ in × 4 in square (3·175 mm × 10 cm square) plaque fed with tab gate area 0·022 in^2 (0·1492 cm^2).

Moulding B: $\frac{1}{8}$ in × 2·3 in square (3·175 mm × 5·84 cm square) plaque fed with flash gate [full width of moulding and cross-sectional area 0·02 in^2 (0·129 cm^2)].

U.F. injection grade material.

	Moulding A	*Moulding B*
Shrinkage at right angle to gate (%)	0·84	1·0
Shrinkage in line with gate (%)	1·68	1·04

Generally speaking, the mould shrinkage of most good-quality, injection grade thermosets is very similar to that for the equivalent compression grades. DMC is in a quite different category in so far as accurate and distortion-free mouldings can easily be made under a wide variety of conditions. The glass fibre reinforcement and the very low mould shrinkage of DMC (grades are available with no shrinkage at all) is responsible for this good performance. Perhaps for this reason alone, DMC is worthy of considerable attention as being probably the most technically satisfactory injection material in the thermoset range.

11.3 MATERIAL REQUIREMENTS IMPOSED BY THE PROCESS

The thermoplastic injection moulding process is one where material is converted to a moulded part by the simple physical process of melting, transfer from the barrel to the cavity and then freezing; there is no time-dependent chemical process involved. The thermoset injection process on the other hand, whilst using almost the same equipment, is subject to both a physical and a chemical process. The physical process of melting the material in the barrel demands the use of as high a temperature as possible in order to reduce the viscosity of the material to the point where it can be easily injected. However, raising the temperature initiates the chemical process of

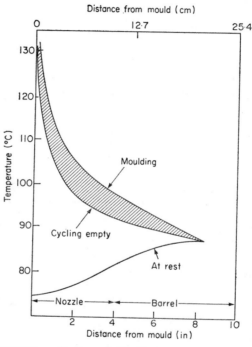

Fig. 11.7. Temperature profile through barrel and nozzle of thermoset injection moulding machine.

curing the substance and thus progressively increases its viscosity. There is thus a competition between these two processes and the temperature at which one chooses to operate the barrel is very much a compromise between achieving the maximum plasticity whilst maintaining an adequately long flow life.

Good injection grade thermosets should therefore fulfil two requirements. They should have low viscosity at moderate barrel temperatures and they should not precure rapidly either at the barrel temperature or at the more elevated temperatures which are momentarily experienced during the transfer to the mould through the runner system.

Figure 11.7 illustrates how, even when the nozzle of the injection barrel is not directly heated, the temperature there can rise substantially above the level of the barrel temperature. This is due not only to conduction from the mould, when the tip of the nozzle is in contact with it, but is also due to frictional heat build-up as the material is injected.

Compression grade thermoset materials selected, for injection moulding, on the basis of their ease of flow rarely give entirely

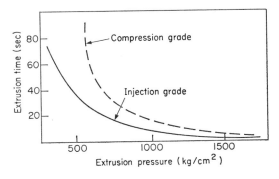

FIG. 11.8. Urea-formaldehyde moulding material—flow characteristics.

satisfactory results because they are too sensitive to these temperatures and too rapidly stiffened due to precure. Even if apparently satisfactory parts can be made, it is usually found that these are highly stressed and subject to cracking on ageing.

Figures 11.8–11.10 characterise these important differences between injection and compression grades of, in this example, urea moulding material. Figure 11.8 shows the relationship between pressure and the time to extrude a fixed quantity of the material through a standard orifice. Although both materials were classified as easy flow on conventional compression flow tests, it is clear from this graph that the compression grade is reluctant to flow at modest pressures and will therefore be less tolerant in the injection moulding process. This observation is confirmed by results which show the influence of pressure on the residual flow of the material (see Fig. 11.9).

These results were obtained in a similar way to those in Fig. 11.8, except that after extruding through the nozzle the flow of the material was measured and compared with the original flow results (i.e. residual flow percentage). It can be clearly seen from Fig. 11.9 that

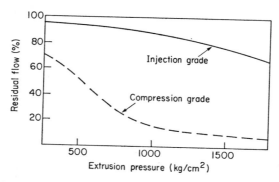

Fig. 11.9. Urea-formaldehyde moulding material—flow characteristics.

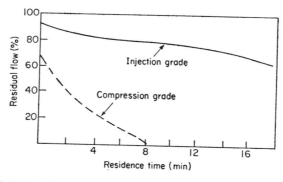

Fig. 11.10. Urea-formaldehyde moulding material—flow characteristics.

increasing extrusion pressure has only a small effect on the injection grade, whereas the compression grade undergoes rapid loss of flow even at modest pressures. In practical terms, therefore, this would mean that while the compression grade might be used on small simple mouldings using low injection pressures and slow injection speeds, any real restriction to the flow which requires higher injection pressures would cause the material to precure and thus lead either to complete failure or the production of poor quality mouldings.

Figure 11.10 further emphasises the difference between the injection and compression grades of material, by showing the effect of increasing residence time in the barrel. It can be seen that the compression grade of material fails to retain an adequate level of flow even for fast cycling conditions of moulding.

Whilst all these data are based on the comparative behaviour of compression and injection grades of urea-formaldehyde moulding material, they apply similarly to most other thermosetting materials. The one important exception is DMC which, by comparison, is exceedingly easy flowing and can be moulded at very low pressures even without barrel heating. No problem is experienced with loss of flow of this material due to precure, as the barrel temperatures normally used are so low (e.g. 20–60°C).

11.4 OPTIMUM MOULDING CONDITIONS

It will be clear from what has already been said that temperatures are often quite critical, particularly in the barrel. It is as important not to use too low a barrel temperature as it is not to use an excessively high one. Low barrel temperatures fail to plasticise the material adequately and thus increase friction in the barrel, causing excessively high local temperatures. The aim should always be to use a barrel temperature which is as high as possible consistent with maintaining an adequate barrel life. Similarly, mould temperatures will be made as high as possible so as to achieve the fastest production rate (provided that this is consistent with the need to maintain good flow of the material through the feed system).

The thermal conditions in the barrel are also influenced by the choice of screw speed and pressure, because the frictional heat developed will increase markedly with rising back pressure and screw speed. The pressures and speeds chosen will vary between different makes of machine but it is always true that the optimum pressure and speed is the lowest value that will still feed the material along the barrel at a rate adequate to complete this within the cure time. This is even more true of DMC where high screw speeds and back pressures cause excessive damage to the glass fibre. (This topic is further discussed later in the paper.)

Injection speed very much influences the cure time of the part because of the frictional heat developed, particularly on passage of the material through the gate. It will always be desirable, therefore, to use the fastest possible speed consistent with avoiding surface porosity. This is a fault which is very often corrected by slowing the injection speed but which is better avoided by providing adequate venting in the tool.

Taking all these considerations into account, optimum conditions for the operation of thermoset injection processes are normally within the range shown in Table 11.4.

TABLE 11.4

NORMAL INJECTION MOULDING PROCESSING CONDITIONS

	General Thermosets	DMC
Barrel temperature	80–110°C	20–60°C
Mould temperature	140–180°C	—
Injection pressure	18 000–28 000 lbf in^{-2} (125–195 MN m^{-2})	5 000 lbf in^{-2} (35 MN m^{-2}) upwards
Hold on pressure	about 6 000 lbf in^{-2} (42 MN m^{-2})	—
Screw speed	25–75 rev/min	10–50 rev/min[a]
Back pressure	800 lbf in^{-2} (5·5 MN m^{-2})	200–800 lbf in^{-2}[a] (1·4–5·5 MN m^{-2})

[a] These parameters are kept as low as possible to minimise fibre breakdown.

11.4.1 Mould Design

Mould design generally follows the same basic rules as those for injection moulding of thermoplastics. For example, it is desirable to feed into the cavity at the thickest section of the moulding: the effect of gate position on distortion will be very similar also. On the other hand, the gate and runner system cannot be as small as those sometimes used for thermoplastic materials, due to the highest viscosities involved and the risk of precuring the material. Notwithstanding this consideration, it is not wise to be overcautious on runner and gate cross-sections because, as described earlier, the wastage of material in the feed system may negate the economic advantage of fast cure. It is generally much more desirable to use modestly small runners and gates but to take care to design these without rapid changes of direction, or any other obstruction, which would impede flow. This means using full round runners, carrying sub-runners through a large radius (rather than at a right-angle from the main runner) and sculpturing the change from runner to gate so as to avoid a hard shoulder which will greatly resist the flow of the material. The effect of such streamlining on the feed system is infinitely more beneficial than increasing the feed cross-sections.

It should also be borne in mind that when trials indicate that there is too much resistance to flow, then a small increase in the gate size is almost always more beneficial than increasing the size of the rest of the feed system.

Moulds with a large number of impressions may require different gate sizes on the more extreme cavities so as to compensate for a pressure drop that may occur at the end of a long main runner. While this can be done successfully by trial and error it is more certain and more elegant to arrange the geometry of the cavities and the feed system in such a way that the distance from sprue to cavity is the same for every impression.

The geometry of the sprue should be dictated by the orifice size of the barrel nozzle, and the sprue size should be marginally larger than this to prevent any resistance to flow. It should have an overall taper of about 2%. A sprue puller on the end is essential and a Z-type is normally preferred.

Gate size should be as large as possible consistent with finishing requirements and available pressure. Tab gates are the most commonly used and are satisfactory in very many cases. They should, however, be placed so that unrestricted jetting into the cavity is avoided (see Fig. 11.5). Wide flash gates are very desirable on parts which are otherwise liable to warp, e.g. flat components (see Fig. 11.6). Replaceable gate pads should be used where possible so that they may be replaced when excessive wear occurs. Generally speaking, most injection grade thermoset materials do not cause excessive gate wear and, perhaps surprisingly, glass reinforced DMC gives very little wear at all, probably because of the very low viscosity of this material.

All moulds should be equipped with an adequate means of venting the cavity of entrapped air. Failure to do this will result in porous mouldings. The natural and most usual position for venting is directly opposite the gate on the parting face of the mould. Such a vent should usually be 0·0015–0·005 in (0·038–0·127 mm) deep and about 0·125–0·5 in (3·175–12·7 mm) wide, depending upon the size of the moulding (Fig. 11.11). Local venting may also be necessary on bosses or other blind parts of the cavity and, in such positions, an ejector pin provides a very convenient method of venting. It should be borne in mind, however, that the pin should be adequately relieved to allow any material flashing down it to escape (Fig.11.12).

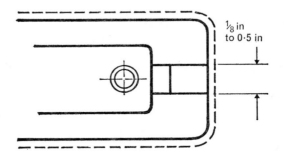

Fig. 11.11. Mould venting on the flash line.

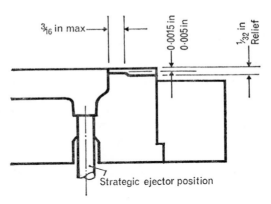

Fig. 11.12. Venting on a boss with an ejector pin.

It may not be possible to use ejector pins as vents on some intricate moulds where air trapping in blind recesses is a problem. While some amelioration of this fault is possible by slowing down the injection speed, this is not always completely successful, and in any case the cure time is thereby extended. A better and highly effective solution to the difficulty can be achieved by applying vacuum to the cavity immediately before injecting the shot of material. Even a modest vacuum (say half an atmosphere or better) will sufficiently remove the entrapped air to allow the use of fast injection speeds and consequent fast cures.

11.5 DEVELOPMENTS

11.5.1 Warm sprue and warm runner moulds

Mention was made, in the earlier part of this paper, of the fact that having to discard the sprue and runners makes a cost disadvantage which in the case of small components may make the process less

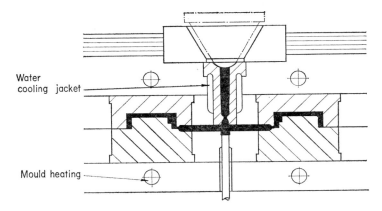

FIG. 11.13. Direct screw transfer, DST warm sprue method.

economic than compression moulding (particularly with more expensive raw materials). Furthermore, as the sprue is often the thickest part to be moulded, it dictates the minimum cure time which needs to be given. Thus, elimination of the sprue, or indeed the whole of the feed system, can substantially improve the economics of the process, both in terms of material and time saving.

Avoidance of such wastage can be achieved by maintaining part, or whole, of the feed system at such a temperature that the material in it remains plastic during the normal curing cycle. This material is then injected into the mould on the next shot.

Figure 11.13 shows an example of this, where the sprue bush is separately maintained at a lower temperature than that of the mould.

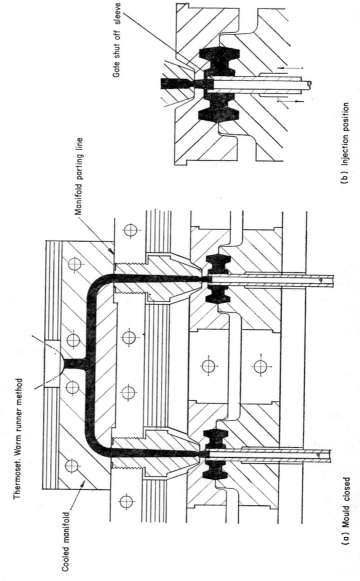

Fig. 11.14. The warm runner method of thermoset injection moulding as applied to the production of a small castor wheel.

The correct temperature to use depends on the particular material and the barrel temperature which is being used on that particular job. It should be somewhat in excess of the barrel temperature in order to avoid stiffening of the flow. For urea (using a barrel temperature of about 95°C), a sprue temperature of about 120°C is usually satisfactory.

While this warm sprue method gives useful economies, and is certainly less difficult than a complete warm runner system, the elimination of all wastage of sprues and runners can be achieved using the warm runner method. The system, but not the temperatures, is very similar to that used for hot runner moulds on thermoplastics. A three-plate mould construction is required to contain a manifold system which distributes the plasticised material to each cavity, or in some instances a collection of cavities. Again, the manifold is maintained at a temperature intermediate between that of the barrel and the mould and the same rules apply as for warm sprue moulds. Because of the long flow path involved, a great deal is demanded of the material in terms of its flow characteristics. Not only does it have to flow a longer distance but it also needs to retain a high degree of flow for more than one cycle of the press.

Not all thermosetting materials may be available in grades adequate to satisfy these exacting requirements.

The difficulties imposed by warm runner moulding dictate the need to give a very careful consideration to the design of the runner system so as to avoid any restriction of flow whatsoever. Full round cross-sections must be used and the radii on all bends must be substantial.

Despite these difficulties it is possible to operate very successfully the warm runner method on suitable moulds. Figure 11.14 shows an example of a small castor wheel where this was done and where the mouldings were also automatically degated. This is achieved by operating a sleeve ejector immediately after filling the cavity and so cutting off the moulding from the feed system—see Fig. 11.14(a). Apart from saving on finishing costs, this method allows the use of large gates without disadvantage.

The number of branches used on the warm runner manifold will be decided on the basis of the number of impressions to be moulded and the tolerance of the material to be used. It is unusual to exceed four branches and this means that, in moulds with many impressions, the cavities will be arranged in such a way that each branch of the manifold will feed a small collection of cavities. In this instance there

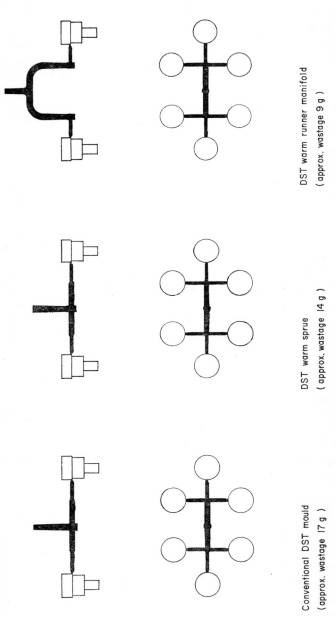

Fig. 11.15. Reduction in wastage by use of warm runner and warm sprue systems.

will be a small wastage of material for each of the sub-runner systems. However, these will usually be quite small in proportion to the whole of the feed system.

Figure 11.15 shows the comparative saving that can be achieved with warm runner and warm sprue moulds as compared with a conventional system.

11.5.2 Injection-compression moulding

A common disadvantage of standard thermoset injection moulding may be the appearance of unsightly flow marks on the surface of mouldings. This can occur when the flow of material is divided by a core pin or some other interruption to the flow. The problem can be very effectively avoided by using what has come to be described as the injection-compression technique. This involves injecting the material into a partially open mould and then, when the shot is completely dispensed, the mould is closed by the compression stroke provided by the clamp. The ability to do this usually requires some modification of the standard injection moulding machine but this is often available on more up-to-date presses. The injection-compression technique, of course, requires a slightly different mould construction as a cavity is required to contain the material. A typical design is shown in Fig. 11.16.

Experience with this technique is fairly limited so far but it appears to be very versatile. Apart from the improvement in surface finish that can be obtained, it is possible to produce much larger mouldings than by conventional means because there is very much less restriction to the flow of the material into the cavity. The 'gate' is effectively very large indeed when the mould is in the open position, whereas when the mould is finally closed it can be arranged that the gate is shut off altogether and a finishing operation thus avoided.

An important disadvantage of the process is that cure times tend to be longer because the considerable frictional heat built up when the material passes through the gate with the usual injection system is reduced.

11.5.3 Injection moulding of DMC

The form of DMC, a fibrous dough, places this material apart from other thermosetting plastics for injection moulding. It is very much easier flowing than all other thermosetting materials and requires special feeding arrangements on the injection machine. It is only

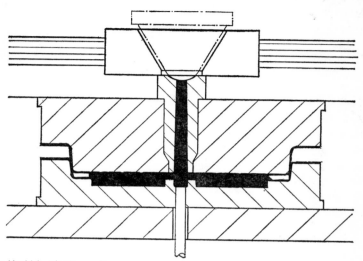

(a) Mould in injection position

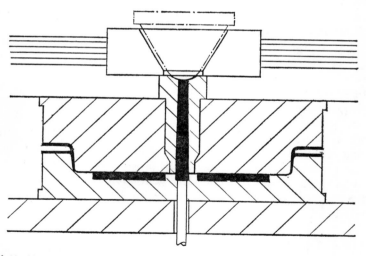

(b) Mould closed

Fig. 11.16. Direct screw transfer, DST injection-compression method.

fairly recently that such machines have become available but there is now rapidly growing activity in this field.

The use of DMC on conventional thermosetting injection machines is usually accommodated by substituting a pressurised stuffing hopper for the normal gravity-fed arrangement. This is usually no more than just a cylinder and piston which forces the DMC from the hopper into the start of the screw (Fig. 11.17).

Although some machines are available which use a plunger rather than a screw for conveying the material along the barrel, it is common

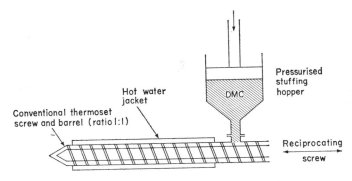

FIG. 11.17. Feeding arrangement for DMC injection moulding.

...ne ways, satisfactory to use the standard thermoset screw... first believed that the action of the screw would cause excess damage to the fibre reinforcement in a material of this kind. This appears to be less of a danger than might be supposed. This is to say that damage of the fibre and consequent loss of strength does not occur when injection moulding DMC. However, it is usually found that most of this loss occurs on passing through the gate because this is the most restricted part of the system. The precise level of strength in the final moulding depends on very many factors. For example, the glass content and flow of the DMC, machine size, screw speed (Fig. 11.18), back pressure, injection speed, barrel temperature, dimensions of the runner system and the size and geometry of the gate. Some typical figures for the retention of strength obtained after passing the material through the machine and into the mould are given in Table 11.5. These results clearly indicate the

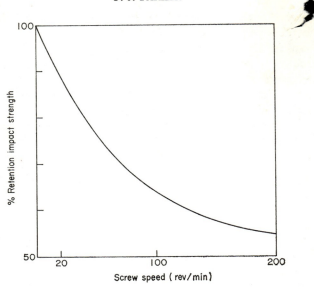

Fig. 11.18. Effect of screw speed on strength of DMC.

TABLE 11.5

THE EFFECT OF THE SCREW INJECTION PROCESS ON THE RETENTION OF STRENGTH OF DMC

	DMC passed through:		
	barrel and screw (no nozzle)	barrel, screw and nozzle	screw, nozzle and... barrel... nner and
Retention of impact strength	65–85%	60–80%	35–?0%[a]
Retention of flexural strength	—	—	60–90%[a]

[a] The lower result is obtained with small gates, the higher with direct sprue feeds.

Note—The results above are obtained by compression moulding the DMC after passing through the parts of the machine and tool indicated in the table. The values are expressed as a percentage of the original compression moulded strength.

very considerable effect that the gate can have on the strength of the moulding, whereas the fall in strength in the barrel is not excessive. It is also worth noting that the effect on flexural strength is not nearly as great as the effect on impact strength. This is because the former property is not as sensitive to fibrillation of the bundles of fibres as is impact strength.

Perhaps the greatest value of the injection moulding process for DMC is that it makes it possible to produce economically large numbers of small parts from this material, whereas to attempt to do this by compression moulding would be very tedious and expensive. It is likely, therefore, that this will very substantially expand the application of the material into fields where, before, the more expensive, but convenient to handle, granular alkyds or DAPs had to be used.

11.6 CONCLUSIONS

The injection moulding process for thermoset materials has only been available, and in commercial use, for a few years but the tremendous economic advantage of this process has led to a very rapid growth in its use and to the development of the technology in terms not only of machines but also of mould design and the materials themselves. While it still remains a minor fabrication process for thermosets, compared with compression moulding, there are indications that the use of the process will grow to embrace a very substantial proportion of the total moulding production of thermosets. The economies of the process are also likely to lead to an expansion of the application of thermosets as they become more competitive with many of the thermoplastic materials.

ACKNOWLEDGEMENTS

Table 11.2 and Figures 11.1, 11.2, 11.4, 11.5, 11.6, 11.11 and 11.12 are reproduced by courtesy of *British Plastics*.

Index

Abrasion resistance, 47, 87, 94
ABS copolymer, 82, 105
Accelerators, 44, 52
Acetal resins, 150
Acetyl acetone peroxide, 34
Acid anhydrides, 112
Acrylate copolymers, 36
Acrylic resins, 14, 25, 29, 35, 46, 113, 114
Additives, 4, 5, 6, 7, 36, 37, 40, 48
Adhesion, 87, 113, 131
Adhesives, 14, 17, 18, 78
Aerospace industry, 42
Aircraft industry, 42
Alaskan oilfield, 125
Alcohols, 43
Alkoxy groups, 134
Alkyd resins, 2, 3, 5, 7, 8, 11, 14, 25, 29, 163, 167
Alkyl radicals, 133
Alkylamine radicals, 133
Alkylation, 28
Aluminium
  chloride, 41
  hydrate, 40
  ions, 128
Amides, 133
Amine/cobalt ratio, 33
Amines, 112, 113
Amino resins, 2–5, 13–30, 42
Aminoplasts, 2–5, 13–30, 42
Aminosilane, 52
Ammonia, 15, 44
Antifade characteristics, 48
Antimony oxide, 37, 39
Anti-static agents, 133, 136
Arab oil policy, 9
Aralkyl ethers, 43, 44, 52
Arc welding, 51
Arcing, 47
Aromatics, 41–45
Arrestor beds, 19
Artificial snow, 19
Asbestos, 16, 45–50
ASTM
  media, 63, 64
  methods, 40, 54, 61, 62, 64
Autoclaves, 15
Automotive industry/products, *see* Motor industry
Azomethine groups, 44

Bagasse, 58, 59
Baking equipment, 51
Barcol hardness, 61, 62
BASF, 16
Bearings, 51
Bedding, 4, 6, 85

189

Benelux, 89
Benzene, 41
Benzoquinone, 31, 32
Benzoyl peroxide, 31, 32
Benzyl chloride, 41, 42
Benzyl trimethyl ammonium chloride, 32
BF_3-amine adducts, 112, 113
Bisphenol A, 112, 113
Blowing agents, 40, 77, 78, 79
BMC, see Bulk moulding compounds
Boards, 17, 46, 53, 54, 55
Boat hulls, 144
BOC (British Oxygen Company), 16
Body fillers, 5
Bonding, 5, 7, 52
  glass/matrix, 130–136
Brake lining, 47, 48
Bromine, 39
BS methods/specifications, 37, 54, 68, 69
BTMAC, see Benzyl trimethyl ammonium chloride
Building industry, 4–8
Bulk moulding compounds, 33, 34, 36
  storage, 97
Bumpers, 77, 82–85
Burning properties, 37
Butanol, 26, 27, 28
Butoxy content, 27
Butyl catechol, 31, 32
Butyl styrene, 35
Butylated amino resins, 25, 26

Calcium ions, 128
Capacitors, 118
Caprolactam, 113, 114
Carbon
  dioxide, 15
  fibre, 43, 52
  monoxide, 15
Carbon-graphite binding, 60

Carcinogenicity, 24, 80
Carnauba wax, 71
Cars, 17, 25
Case histories (furanes), 72–74
Catalyst/accelerator/inhibitor systems, 31–33
Catalyst injection, 139
Catalysts, 15, 20–29, 41, 43, 58–61, 64, 71, 77, 78, 139
Catalytic dehydrogenation, 15
Cavity walls, 17, 18, 19
Cellular polyurethanes, 96–109
Cellulose
  acetate butyrate, 36
  esters/ethers, 14
  fabric, 20, 24
  filler, 16, 17
Chain extenders, 112
Chairshells, 85, 86
Char, 40
Chemical plant, 5
Chemical resistance, 45, 50, 51, 58, 60, 64, 65, 71, 72, 112, 113, 124
Chipboard, 7
Chloramines, 22, 23
Chlorendic acid, 37, 38
Chlorinated paraffins, 37, 39
Chlorine, 37, 38
  damage, 24
  retention, 23, 24, 25
Chloromethyl compounds, 42, 43
Chopped strand mat, 137
Chromium compounds, 131, 132
Cladding, 4
Clamping units, 151–154
Closures, 16
Cloud chamber, 123
Coated fabrics, 86–88
Coating
  powders, 110–114
  resins, 14, 25–29
  techniques, 117–126
Coatings, 4, 5, 6, 7, 78, 86, 87, 88, 110–114
Cobalt
  octoate, 33
  soap, 31, 32, 33

INDEX

Cold cure foam, 80, 81, 82, 104
Cold press moulding, 143, 144
Commutators, 50
Composites, 43, 52–56, 60, 72
Compression moulding, 48, 50, 162–166, 173, 174
Consumer durables, 4, 36
Consumption
  plastics, 2, 3, 4, 7, 8, 9
  polyurethanes, 4, 76
  powders, 110, 125
  thermosets, 2, 3, 8
  see also Usage
Contact moulding, 137
Continuous roving, 137, 138
Continuous strand mat, 137, 138
Conveyors, 103–106
Cooking utensils, 51
Copying machines, 51
Corn cobs, 58, 59
Corona charging, 119
Corrosion resistance, 4, 58, 64, 66, 72, 74, 125
Cotton fabrics, 21, 22
Cottonseed hulls, 59
Coupling agents, 52, 136
Creasing, 20–25
Creep, 131
Cresylics, 2, 3, 14
Cross-linking, 23, 24, 25, 29, 31, 42–45, 48, 52, 59, 60, 77, 78, 79, 112, 113, 145
CSM, see Chopped strand mat
Cure time, 32, 33
Curing, 81, 82, 104, 111, 112, 143
Cyclohexanone peroxide, 34

DAP, see Diallyl phthalate
DBNPG, see Dibromoneopentyl glycol
DDM, see Diamino diphenyl methane
Deep freeze, 4, 6
Density, 19, 90
Dialkyl phthalate, 35, 163, 167, 187
Diamino diphenyl methane, 112

Diamino propane, 24
Dibromoneopentyl glycol, 37, 38, 39
Dichloroxylene, 42, 43
Dicyandiamide, 16, 112, 113
Dielectric properties, see Electrical properties
Dielectric strength, 55, 62
Diesel engines/fuels/cubes, 47, 50
Diethyl aniline, 33
Differential thermal analysis, 65
Difuryl ether, 60
Difuryl methane, 60
Dihydroxyethylene urea, 24
Dimedone, 25
Dimensional stability, 45, 48, 51
Dimer acids, 112
Dimethoxyxylene, 43, 44
Dimethylene amino linkage, 44
Dimethylol
  derivatives, 24, 25
  dihydroxyethylene urea, 23, 24, 25
DIN tests, 61, 62
Diphenyl ether, 42
Diphenyl methane diisocyanate, 4, 104
Direct screw transfer, 179, 183, 184
Distortion, 169, 170
DMC, see Dough moulding compounds
Domestic appliances, 16, 51
Dough moulding compounds, 163, 167, 169, 171, 175, 176, 183–187
Drip dry, 21
Drostholm process, 142
Dry-cleaning resistance, 21, 87
DSM, 16
DTA, see Differential thermal analysis
Durable press process, 24, 25
Dyes, 21

EFTA, 89
Elastomers, 78
Electric motors, 43, 47

Electric strength, 55, 62
Electrical industry, 4, 5, 6, 8, 16, 46, 47, 55
Electrical properties, 47, 48, 50, 52, 54, 55, 56, 62, 120
Electronic components, 118
Electrophoresis, 125
Electrostatic coating, 111, 118–125
Energy absorption, 82, 83, 85
Engine lubricating oil, 45, 51, 90
EP, see Epoxide resins
Epichlorhydrin, 112
Epoxide
  boards, 53, 54, 55
  laminates, 54, 55
  powders, 124
  resins, 2–8, 29, 42, 44, 46, 47, 51–56, 111, 112, 113, 132, 142, 144
Epoxy radicals, 133
Epoxy resins, see Epoxide resins
Epoxy silane, 52
EPS, see Expanded polystyrene
Ethyl carbamate, 24
Ethyl hexanol, 28
Ethyl methyl ketone, 52
Ethylene diamine, 23
Ethylene urea, 23, 24
Europe, 89, 93, 125
Exotherm curves, 34
Expanded polystyrene, 4
Extruders, 116

Fabric coating, 86–88
Faraday cage effect, 121
Fatigue resistance, 81
Ferric chloride, 43
Fertiliser, 111
FHP, see Fractional horsepower
Fibreglass reinforced plastics, 127–148
Fibre/matrix bonding, 146, 147
Filament winding, 140, 141, 142, 146
Fillers, 36, 37, 48–51
Film forming, 25, 132, 133, 136
Finance, 11

Fire
  hazards, 31, 37–40, 55, 118
  propagation test, 68, 69
  resistance, 68–70
  retardants, 37, 70, 72,
Flame
  proofing, 21
  spraying, 123
Flammability, 81
Flexibility, 89, 91, 94
Flexible foam, 4, 6, 77, 78, 81, 85, 86, 93, 105, 106
Flexural strength, 45, 46, 48, 49, 53, 62–67, 132, 146, 186
Flock spraying, 123
Flooring, 6
Flow
  characteristics, 172, 173, 174
  marks, 169, 170
Flowback, 155, 156, 157
Flower arrangement, 19
Fluidised bed, 111, 117, 118, 123, 124
Fluon, see Polytetrafluoroethylene
Fluorocarbon, 79
Foam, 4, 6, 14, 17, 18, 19, 76–81, 85–89, 91, 93, 106
  density, 19, 89
  gun, 18
  moulding, 104, 105, 106, 107
  rigid, 4, 6, 76, 77, 78, 85–89, 106
  semi-rigid, 78, 79, 82
Foamex process, 106, 107
Footwear, 77, 79, 87, 89–93
Formaldehyde, 15, 18, 23, 24, 26, 27, 28
Formalin, see Formaldehyde
Forming size, 130–133
Foundry industry, 5, 60, 68
Fractional horsepower motors, 123
France, 15, 89
Friction
  linings, 47
  materials, 5
  properties, 51
Friedel–Crafts resins, 41–57
FRP, see Fibreglass reinforced plastics

INDEX

Furane resins, 58–75
Furfural, 58, 59
Furfuryl alcohol, 58–61
Furniture, 4, 6, 7, 8, 36, 85–88

Garments, 24, 25, 87
Gas
  analysis, 65
  reservoirs, 47
Gel time, 32, 33, 71
Germany, 15, 17, 89, 94, 111
Glass
  cloth, 46, 52–56, 135
  fibre, 37, 58–71, 74, 127–148, 171
  filler, 16
  matrix bonding, 130–136
  reinforcement, 45–49, 58–71, 74
  technology, 128–130
Glycidal groups, 114
Glyoxal, 24
Graphite, 51, 71
Gross national product, 9, 10
Growth rate, 2, 9, 10, 13

Halogens, 37, 38, 39, 134
Hand lay-up, 70, 71, 137, 138, 139, 145, 147
Handles, 51
Hardness, 61, 62
Heat resistance, 42, 48, 112, 113, 131
Heaters, 47
Hexamethoxymethyl melamine, 113
Hexamethyl ether of hexamethylol melamine, 25, 28, 29
Hexamethylene tetramine, 44
Hexamine-cured resins, 44–51, 55
High temperature strength, 45, 47, 52, 53, 65–68
  see also Mechanical strength
HMEHMM, *see* Hexamethyl ether of hexamethylol melamine
Hot-water pumps, 51
Household goods, 6
Houses, 17

Hydraulic oils, 45
Hydrogen, 15
Hydrogen chloride, 42, 43
Hydrophilic foam, 93
Hydroquinone, 32
Hydroxyl compounds, 77, 78
Hypochlorite bleach, 22
Hysteresis, 81, 82

ICI, 15
Imino groups, 22, 25
Inhibitors, 31, 32, 33
Injection-compression moulding, 183, 184
Injection moulding, 16, 48, 50, 51, 149–187
Innovation, 9, 10, 11
Insulation
  electrical, 43, 46, 47, 50, 55, 56
  thermal, 5, 6, 7, 8, 17, 19, 87, 90, 94
Integral skin foams, 78, 79, 80, 86, 91
In-the-skin moulding, 107
Intumescent coatings, 40
Iporka, 17
Irons, 51
Isobutanol, 28
Isocyanates, 77, 78, 96, 113
Isophorone diisocyanate, 113
Italy, 15, 89

Japan, 15, 114

Ko-Kneader, 17, 116
Koratron process, 25

Laminates
  fire protection, 40
  manufacture, 17, 33, 139, 140
  performance data, 145, 146
  properties, 50, 61–70
  reinforced, 37, 45–48, 52
  resins for, 4, 5, 6, 14, 61–72, 127

Laundering, 21, 22, 24
Lavatory seats, 16
Leather, 89
Limiting oxygen index, 37, 38, 39, 54, 70
Linking agents, 131–136
  see also Cross-linking
Load/deflection curves, 82, 83
LOI, see Limiting oxygen index
Lorry cabs, 144
Loss tangent, 45, 52, 54, 55
Low-profile resin system, 33–36
Lubricants, 133, 136

Machinability, 45
Machinery
  injection moulding, 149–161
  shoe soles, for, 92, 93
Magnesium, 23, 24, 128
Manpower, 11
Marble bushing, 130
Marine use, 4, 5
Marketing, 10
MDI, see Diphenylmethane diisocyanate
Mechanical strength, 17, 42–56, 62, 67, 94, 112, 113, 131, 146, 186
MEKP, see Methyl ethyl ketone peroxide
Melamine, 15, 16, 26–29
  boards, 54
  formaldehyde resins, 1, 2, 7, 8, 16, 17, 21, 22, 29, 163, 167, 168
Melinex, 71
Mercapto radicals, 133
Metering, 97–101
Methacrylato chromic chloride, 133, 134
Methacryloxy radicals, 133
Methanol, 15, 24, 28, 29, 43
Methoxy methyl uron, 24
Methyl carbamate, 24
Methyl chloride, 41
Methyl ethyl ketone peroxide, 31, 32, 34

Methyl methacrylate, 35
Methylene groups, 44
Methylol
  derivatives, 23
  groups, 27
  melamines, 20
  stearamide, 21
  urea, 20
MF, see Melamine-formaldehyde resins
Mica, 48, 49, 52, 55, 56
Michigan Chemical Oxygen Index, 70
Microcellular polyurethane, 78, 82, 83
Mixing, 101–103
Monochlorostyrene, 35
Monomers, 34, 35
Mononuclear aromatics, 43
Mortar, 58, 60
Motor
  bodies, 36
  components, 51
  industry, 4, 7, 8, 77, 79, 86, 125
Mouldings, 34, 36, 48, 50, 76, 81, 86, 107, 139
  compounds, 33, 36, 45, 48–51
  materials, 5, 14, 16, 17
  powders, 4, 13, 14, 48
  techniques, 48, 50
Moulds, 103–106, 160, 161
  design, 176–183
  release agents, 71
Mylar, 71

Naphthalene, 43
National Economic Development Office, 1, 4, 7, 9
National Institute for Economic and Social Research, 10
NEDO, 1, 4, 7, 9
Network modifiers, 128
Neutron magnetic resonance, 25
NIESR, see National Institute for Economic and Social Research
Nitrogen, 27

NMR, *see* Neutron magnetic resonance
Nylon, 21, 111, 150

Oat hulls, 59
OECD, *see* Organisation for Economic Cooperation and Development
Oil resistance, 45, 51, 90
Operating ratio/physical property graphs, 92
Organisation for Economic Cooperation and Development, 10
Organometallic compounds, 44
Overspray, 121, 124
Oxygen index, *see* Limiting oxygen index

Packaging, 4, 6, 8, 93
Paint, 5, 114
  spraying, 122, 124, 125
Paper, 4, 14, 17
Particle board, 17
Patterns of usage, 1–13
Peak temperature, 32, 33
Peel strength, 56
Pentabromotoluene, 37, 39
Permittivity, 45, 52, 54, 55
Peroxides, 31–34
Petrol tank safety, 85
PF, *see* Phenol-formaldehyde resins
Phenol, 44, 52
Phenol-formaldehyde resins, 1–5, 7, 8, 14, 42, 45, 47–50, 54, 60, 163, 167
Phosphates, 37
Phosphoric acid, 40, 60
Phosphorus, 38, 39
Phthalic anhydride, 27
Pigment pastes, 21
Pipes, pipelines, 5, 74, 107, 108, 125, 141, 142
Plasma torch, 123
Plastics, consumption/production, 1, 2, 3, 4, 7, 8, 9, 13, 14, 15

Plywood, 17
PMDA, *see* Pyromellitic dianhydride
Pollution, 118, 122, 124
Polybenzyl, 41, 42
Polycaprolactones, 36
Polycarbonate, 150
Polyester resins, 2–8, 14, 31–40, 42, 68–72, 90–93, 113, 132, 139, 143, 144
Polyethers, 90–93
Polyethylene, 14, 36, 110
  terephthalate, 71
Polyhydroxy compounds, 40
Polyimide resins, 46
Polyisobutylene, 36
Polymers, 34, 35
Polyolefins, 4, 78, 79, 81, 93, 96
Polyols, 150
Polypropylene, 14
Polystyrene, 4, 36, 150
Polytetrafluoroethylene, 14, 71
Polyurethane, 2, 3, 4, 7, 8, 11, 14, 76–109
  cellular, 78, 82, 83, 96–109
  classification, 78
  consumption, 76, 77
  formation, 77–79
Polyvinyl acetate, 14, 132, 133
Polyvinyl alcohol, 71
Polyvinyl carbazole, 36
Polyvinyl chloride, 14, 87, 89, 107, 111
Polyvinyl ether, 36
Postal bags, 94
Postcure, 50–53, 61
Powder
  coating, 4, 16, 110–114, 124–126
  manufacture, 114–117
  moulding, 4, 13, 14, 48
  processes, 110–126
Prepregs, 52
Pressure bag moulding, 139, 140, 145
Price structure, 12
Printed circuits, 56
Prints, 21
Processing equipment, 96–109

Production, plastics, 1, 2, 3, 7, 13, 16, 31
   costs, injection moulding, 165, 166
Promoters, 31, 32
Propylene urea, 24
PTFE, *see* Polytetrafluoroethylene
PU, *see* Polyurethane
Pumps, 51, 98–101
PVC, *see* Polyvinyl chloride
Pyrolytic products, 65
Pyromellitic dianhydride, 112, 113

Radiation resistance, 45, 113
Raw materials, 15, 96, 97
Reactive groups, 133
Refinery gas, 15
Refractory binders, 68
Refrigerant, 11, 79, 80
Refrigerators, 4, 6, 17, 25, 76, 89
Reinforced laminates, 37, 45–48, 52
Resilience, 90
Resin
   injection, 144, 145
   rubber, 89
Resistance
   abrasion, 47, 87, 94
   chemical, 45, 50, 51, 58, 60, 64, 65, 71, 72, 112, 113, 124
   corrosion, 4, 58, 64, 66, 72, 74, 125
   dry-cleaning, 21, 87
   fatigue, 81
   fire, 68–70
   heat, 42, 48, 112, 113, 131
   oil, 90
   radiation, 45, 113
   slip, 90, 94
   solvent, 51
   wear, 51, 89, 90, 91
Resistivity, 120
Resistors, 118
Rice hulls, 59
Rigid foams, 4, 6, 76, 77, 78, 85–89, 106
Rigidity, 131

Roads, 94
Rot proofing, 21
Roving, 137, 138, 142
Rubber, 89
Running tracks, 94
Runways, 19

Safety car, 85
Sag factor, 81
Sandwich panels, 104
Scandinavia, 94
Seat mouldings, 81
Self-extinguishing, 37
Self-smoothing, 21
Semiconductors, 51
Semi-rigid foams, 78, 79, 82
Sheet moulding, 33, 36
Shelf stability, 52
Shoe soles, 89, 90
Shrinkage, 34, 35, 36, 50, 169, 170, 171
Silanes, 71, 131–136
Silanol groups, 133, 136
Silica, 48, 49, 51, 128
Silicon compounds, 131, 132
Silicone
   boards, 53, 55
   laminates, 53, 54, 55
   resins, 42, 46, 51, 54, 71, 77, 78, 133
Siloxanes, 136
Slabstock, 76, 104, 105, 106
Slip resistance, 90, 94
Slot wedges, 47
SM, *see* Continuous strand mat
SMC, *see* Sheet moulding compounds
Smoke, 40, 68, 70, 72
Smooth drying, 21–24
Sodium ions, 128
Solvent resistance, 51
Soviet Union, 15
Spray
   drying, 116
   guns, 119–122
Spray-up, 70, 71, 139, 145, 146

INDEX

Spread of flame test, 37, 68, 69, 70
Stability
  dimensional, 45, 48, 51
  shelf, 52
  thermal, 42–53, 56, 65
Stannic oxide, 42
Starch, 132
Stearamide, 21
Stearates, 71
Steiner tunnel test, 69
Stiffening, 21
Storage tanks, 72, 73, 74, 139, 141, 142
Stoving enamels, 17, 25
Strength retention, 45–50, 131, 186
Strip burning test, 37
Stripping time, 71
Structure, 11
Styrene, 14, 35, 36, 71
Surface finish, 112, 113
Switchgear, 47
Synergists, 37, 39
Synthetic fibres, 21

Tableware, 16
Tanks, 72, 73, 74, 139, 141, 142
TBPA, *see* Tetrabromophthalic anhydride
TCPA, *see* Tetrachlorophthalic anhydride
TDI, *see* Toluene diisocyanate
Teflon, *see* Polytetrafluoroethylene
Tensile strength, 22, 54, 62, 67, 129, 131, 146
  see also Mechanical strength
Tetrabromophthalic anhydride, 37, 38, 39
Tetrachlorophthalic anhydride, 37, 38, 39
Textile
  equipment, 51
  finishing, 17, 20–25
  resins, 4, 6, 14
TGA, *see* Thermal gravimetric analysis
Thermal expansion, 131
Thermal gravimetric analysis, 65

Thermal stability, 42–53, 56, 65
Thermoplastics
  availability, 4
  coating, 111
  development, 1
  injection-moulding, 149, 150, 171
  mouldings, 86
  powders, 110, 111
  production, 1, 11, 14
  reinforced, 127–148
Thermosetting plastics
  coating, 14, 25–29, 110–126
  consumption, 2, 3, 7, 8
  development, 1, 2
  growth rate, 13
  injection-moulding, 149–187
  powder processes, 110–126
  production, 1, 2, 3, 7, 8, 9, 14, 15
  reinforced, 127–148
TMA, *see* Trimellitic anhydride
Toluene, 41, 43
  diisocyanate, 4, 78, 104
  sulphonic acid, 29, 60
Tooling, 6
Toughness, 82
Toxic fumes, 40
Transetherification, 28, 29
Transfer
  coating, 87
  moulding, 48, 50
Transformers, 47
Transport industry, 4, 5, 6, 79, 85
Triethyl phosphate, 37, 38, 39
Trimellitic anhydride, 112
Tris (dibromopropyl) phosphate, 37

UF, *see* Urea-formaldehyde resins
UK
  economy, 9, 10
  industry, 9, 10
  materials shortage, 11, 12
  plastics industry, 9, 10
  powder consumption, 110
  production, plastics, 1, 2, 13–16, 31
  usage, 9, 14, 85, 86, 89, 147

UP (Unsaturated polyesters), *see* Polyesters
Upholstery, 85, 87
Urea, 15, 16, 18, 21, 24, 26, 163, 165, 167, 173
Urea-formaldehyde
 mouldings, 166, 168, 174
 resins, 1, 2, 7, 8, 16–22, 60
Uron, 24, 25
USA, 13, 15, 16, 73, 82, 83, 85, 89, 98, 125, 142, 147
Usage, patterns of, 1–12, 14

Vacuum bag moulding, 140, 141, 145
Vinyl acetate, 35
Vinyl chloride, 35
Vinyl radicals, 133
Vinyl skin, 82
Vinyl toluene, 35
Viscosity/time graph, 91, 92
Volan, 133

Warm runner/sprue moulds, 179–182
Washing machines, 17
Water
 absorption, 45, 54, 62, 91
 proofing, 21
 vapour permeability, 87
Wear resistance, 51, 89, 90, 91
White spirit, 26, 27
Wood, 7, 86
Woven roving, 138
Wrapround, 121

Xylene, 26
Xylok resin, 44

Yachts, 139

Zinc
 carbonate, 24
 ions, 128
 nitrate, 23, 24